普通高等教育"十三五"规划教材

Android移动开发详解
——从基础入门到乐享开发

张传雷 主编
陈亚瑞 于 洋 孙 迪 潘 刚 副主编

电子工业出版社
Publishing House of Electronics Industry
北京·BEIJING

内容简介

本书是一本基于 Android 的移动应用开发教科书，旨在帮助读者快速学习 Android 的基本概念，掌握 Android 的开发技能。

本书分为 3 篇：Android 移动开发概述、Android 基础编程和 Android 高级编程。Android 移动开发概述包括第 1 章～第 3 章，介绍 Android 历史、Android 体系结构及特点、其他主要智能手机开发平台、IDE Eclipse、Android Studio、Kotlin 及搭建 Android 开发环境等。Android 基础编程包括第 4 章～第 7 章，介绍 Android 应用中的基本概念、Android 工程结构、用户界面介绍、Android 颜色的基本用法和介绍、基本组件介绍和应用、获取屏幕属性、Android 图形开发框架、Graphics 类、动画设计、Android 中的文件操作、SharedPreferences、SQLite 数据库数据存储。Android 高级编程包括第 8 章～第 12 章，介绍 Android 多媒体编程、Android 网络与通信编程、Android-OpenGL 应用开发、Android 传感器开发及 Android NDK 开发技术等。本书采用代码驱动式叙述方式，注重代码的讲解。

本书既可作为高等学校计算机、物联网类专业的移动开发技术课程的教材，也可作为各类职业培训机构的 Android 培训教材，还适合作为广大 Android 初学者的参考书。

未经许可，不得以任何方式复制或抄袭本书之部分或全部内容。

版权所有，侵权必究。

图书在版编目（CIP）数据

Android 移动开发详解：从基础入门到乐享开发/张传雷主编. —北京：电子工业出版社，2018.8
ISBN 978-7-121-33892-2

Ⅰ. ①A… Ⅱ. ①张… Ⅲ. ①移动终端－应用程序－程序设计－高等学校－教材 Ⅳ. ①TN929.53

中国版本图书馆 CIP 数据核字（2018）第 055843 号

策划编辑：戴晨辰
责任编辑：谭丽莎
印　　刷：北京虎彩文化传播有限公司
装　　订：北京虎彩文化传播有限公司
出版发行：电子工业出版社
　　　　　北京市海淀区万寿路 173 信箱　邮编：100036
开　　本：787×1092　1/16　印张：14.5　字数：335 千字
版　　次：2018 年 8 月第 1 版
印　　次：2021 年 5 月第 4 次印刷
定　　价：39.00 元

凡所购买电子工业出版社图书有缺损问题，请向购买书店调换。若书店售缺，请与本社发行部联系，联系及邮购电话：（010）88254888，88258888。

质量投诉请发邮件至 zlts@phei.com.cn，盗版侵权举报请发邮件至 dbqq@phei.com.cn。

本书咨询联系方式：dcc@phei.com.cn。

前言

自 2011 年第一季度以来，Google 的移动操作系统 Android 在全球智能手机操作系统市场的份额稳步增加。截至 2016 年第四季度，Android 以 81.7％的市场份额领跑全球市场，苹果的 iOS 操作系统则位居第二。同时，Android 也是全球平板电脑最常用的操作系统，2016 年其全球市场份额为 66％。Android 成功的原因之一是其多种版本的不断改进，每个新版本都提供了更先进的功能、更快的访问互联网的速度或越来越优质的视频和音频。Android 受欢迎的另一个原因是它与移动设备制造商进行强大合作，截至 2017 年，有 85％的新智能手机使用了 Android 操作系统的设备。

Android 市场份额的扩大也带动了基于 Android 的移动软件的发展，越来越多的开发者开始学习 Android 开发，涌入 Android 开发的各个领域。其中，有一些是有软件开发经验的开发者，也有一些是初学者或在校学生。然而，现在市场上的 Android 书籍多以讲解 Android 知识点为主，或者是附上大量的源代码加以介绍。本书的几位作者中，有的具有 Android 开发的实际经验，有的具有丰富的课堂教学经验。因此，本书最大的特点就是在讲解 Android 应用开发各知识点的同时，还分享了很多实际开发经验，这些经验一般都无法系统地从现有的书籍或网络上获得。这些都是作者在开发和教学过程中思考的成果，是作者多年从事软件开发和教学工作的技术沉淀。

总之，本书是一本基于 Android 的移动应用开发教科书，旨在帮助读者快速学习 Android 的基本概念，掌握 Android 的开发技能。本书分为 3 篇：Android 移动开发概述、Android 基础编程和 Android 高级编程，主要采用代码驱动式叙述方式，注重代码的讲解。Java、XML、Linux 等基本知识作为本书的先导性知识，读者应熟悉。

本书包含配套教学资源，读者可登录华信教育资源网（www.hxedu.com.cn）注册后免费下载。

总体来说，对于计算机或物联网工程专业的在校学生，开发一个简单的 Android 应用并不难，但能够深刻了解 Android 开发的基本概念，用最合理的技术开发出一个优秀的 Android 应用并不容易，需要开发者具有丰富的 Android 知识和开发经验。本书既可作为高等学校计算机、物联网类专业的移动开发技术课程的教材，也可作为各类职业培训机构的 Android 培训教材，同时适合作为广大 Android 初学者的参考书。

由于本书涉及知识较多，特别是 Android 每年都有技术的更新和升级，而作者水平有限，很难全部精通，难免有疏漏之处，敬请广大读者批评指正。

最后，感谢为本书提供资料整理和例程测试的同学们，感谢出版社的各位编辑，也感谢家人的理解和支持。

编　者

目录

第1篇 Android 移动开发概述

第1章 Android 简介 ... 3
1.1 Android 历史 ... 3
- 1.1.1 Android 简要介绍 ... 3
- 1.1.2 Android 发展历史 ... 3
- 1.1.3 Android 版本升级 ... 4

1.2 Android 体系结构及特点 ... 7
- 1.2.1 应用程序 ... 9
- 1.2.2 中间件 ... 9
- 1.2.3 硬件抽象层 ... 10
- 1.2.4 操作系统 ... 10

1.3 其他主要智能手机开发平台 ... 10
- 1.3.1 iOS 简介 ... 10
- 1.3.2 Windows CE 简介 ... 11
- 1.3.3 Symbian 简介 ... 11
- 1.3.4 Palm OS 简介 ... 12

第2章 Android 开发基础 ... 13
- 2.1 面向对象编程介绍 ... 13
- 2.2 Android Java 基础 ... 14
- 2.3 XML 基础 ... 15
- 2.4 IDE Eclipse 介绍 ... 16
- 2.5 Android Studio 介绍 ... 16
- 2.6 Kotlin 介绍 ... 17

第3章 搭建 Android 开发环境 ... 19
3.1 Android SDK 介绍 ... 19
- 3.1.1 Android SDK 目录结构 ... 19
- 3.1.2 android.jar 内部结构 ... 20
- 3.1.3 android.bat 批处理常用命令 ... 21

3.1.4 模拟 SD 卡 22
　　3.1.5 Traceview 工具 22
　　3.1.6 ADB 工具 23
3.2 搭建开发环境 24
　　3.2.1 安装 JDK 24
　　3.2.2 安装 Android Studio 24
　　3.2.3 创建 Android 虚拟设备 25
3.3 DDMS 工具 26
　　3.3.1 DDMS 详细功能 27
　　3.3.2 DDMS 工作原理 27
3.4 第一个 Android App 28
　　3.4.1 创建 Hello World App 28
　　3.4.2 Android 工程目录结构 31
　　3.4.3 Android 程序部署与启动 32
　　3.4.4 Android 程序打包安装过程 32
3.5 NDK 开发工具 33
　　3.5.1 NDK 下载 34
　　3.5.2 NDK 开发 34

第 2 篇　Android 基础编程

第 4 章 Android App 基本概念 37
4.1 Android 应用中的基本概念 37
　　4.1.1 Activity 37
　　4.1.2 Intent 41
　　4.1.3 Service 41
　　4.1.4 Broadcast 42
　　4.1.5 Binder 42
　　4.1.6 Permission 42
　　4.1.7 Manifest 43
4.2 Android 工程结构 44

第 5 章 Android 应用用户界面设计 46
5.1 用户界面介绍 46
　　5.1.1 Android 基本布局知识 46
　　5.1.2 View 视图组件 46

目录

- 5.1.3 ViewGroup 视图容器组件 ·············· 47
- 5.1.4 Layout 布局组件及其参数 ·············· 47
- 5.1.5 界面布局 ·············· 48
- 5.1.6 事件处理的简单介绍 ·············· 59
- 5.2 Android 颜色的基本用法和介绍 ·············· 59
- 5.3 基本组件介绍和应用 ·············· 64
  - 5.3.1 Widget 组件 ·············· 64
  - 5.3.2 ListView 列表 ·············· 82
  - 5.3.3 Notification 状态栏提示 ·············· 85
  - 5.3.4 Toast 临时提示框 ·············· 88
  - 5.3.5 Dialog 对话框 ·············· 89
- 5.4 获取屏幕属性 ·············· 92

第 6 章 Android 图形编程 ·············· 94

- 6.1 Android 图形开发框架 ·············· 94
  - 6.1.1 View 类开发框架 ·············· 94
  - 6.1.2 SurfaceView 类开发框架 ·············· 95
- 6.2 Graphics 类 ·············· 97
  - 6.2.1 android.graphics.Color 类 ·············· 97
  - 6.2.2 android.graphics.Paint 类 ·············· 97
  - 6.2.3 绘制几何图形 ·············· 98
  - 6.2.4 android.graphics.Canvas 类 ·············· 98
  - 6.2.5 绘制字符串 ·············· 103
  - 6.2.6 android.graphics.Bitmap 类 ·············· 105
  - 6.2.7 Shade 类 ·············· 115
- 6.3 动画设计 ·············· 117
  - 6.3.1 Tween 动画 ·············· 117
  - 6.3.2 Frame 动画 ·············· 121

第 7 章 Android 数据存储编程 ·············· 122

- 7.1 Android 中的文件操作 ·············· 122
  - 7.1.1 File 类及常用方法 ·············· 122
  - 7.1.2 文件 I/O ·············· 124
- 7.2 SharedPreferences ·············· 127
  - 7.2.1 获取 SharedPreferences 的句柄 ·············· 127
  - 7.2.2 写入共享文件 ·············· 128

7.2.3 读取共享文件 ·· 128
7.3 SQLite 数据库数据存储 ·· 131

第 3 篇　Android 高级编程

第 8 章　Android 多媒体编程 ·· 137
8.1 OpenCore 多媒体架构 ·· 137
8.2 MediaPlayer 编程 ·· 139
  8.2.1 MediaPlayer 主要接口定义 ·· 141
  8.2.2 播放音乐实现 ·· 143
  8.2.3 播放视频实现 ·· 149
8.3 MediaRecoder 编程 ·· 150
8.4 Camera 编程 ·· 155

第 9 章　Android 网络与通信编程 ·· 160
9.1 HTTP 协议原理 ·· 160
  9.1.1 HTTP 简介 ·· 160
  9.1.2 HTTP 的请求报文 ·· 161
  9.1.3 HTTP 的响应报文 ·· 162
  9.1.4 HTTP 的消息报头 ·· 163
9.2 Android 网络编程基础 ·· 164
9.3 HTTP 通信 ·· 165
  9.3.1 HttpURLConnection 接口 ·· 166
  9.3.2 HttpClient 接口 ·· 172
  9.3.3 实时更新 ·· 175
9.4 Socket 通信 ·· 178
  9.4.1 Socket 传输模式 ·· 178
  9.4.2 Android Socket 编程步骤 ·· 180
9.5 Socket 应用 ·· 182
9.6 WebKit 应用 ·· 183
9.7 WiFi 编程 ·· 184
9.8 蓝牙编程 ·· 185

第 10 章　AndroidOpenGL 应用开发 ·· 187
10.1 AndroidOpenGL ES ·· 187
  10.1.1 构建 OpenGL 基本框架 ·· 187

10.1.2　OpenGL 视图显示 ·················· 188
10.2　OpenGL 的三维坐标基础 ·················· 189
10.3　多边形的绘制及其颜色渲染 ·················· 189
10.4　图像旋转 ·················· 193
10.5　3D 三维实体空间 ·················· 194
10.6　映射纹理 ·················· 197
10.7　光照与单击事件 ·················· 201

第 11 章　Android 传感器开发 ·················· 208

11.1　传感器种类 ·················· 208
　　11.1.1　GPS ·················· 208
　　11.1.2　动作传感器 ·················· 208
　　11.1.3　位置传感器 ·················· 209
　　11.1.4　环境传感器 ·················· 209
11.2　GPS 应用 ·················· 209
　　11.2.1　我的位置 ·················· 209
　　11.2.2　更新位置 ·················· 210
　　11.2.3　地图功能 ·················· 212
11.3　Acceleration 传感器 ·················· 213
11.4　Gyroscope 传感器 ·················· 214
11.5　Proximity 传感器 ·················· 214

第 12 章　Android NDK 开发技术 ·················· 216

12.1　NDK 环境的搭建 ·················· 216
12.2　新建 NDK 工程 ·················· 216

参考文献 ·················· 219

第 1 篇
Android 移动开发概述

第 1 章　Android 简介
第 2 章　Android 开发基础
第 3 章　搭建 Android 开发环境

第1章

Android 简介

1.1 Android 历史

随着移动多媒体时代的到来，作为人们必备的移动通信工具，手机开始从通话工具的角色逐渐转向智能化发展。手机凭借其操作系统的开源化和应用软件的多能化，逐渐成为一台迷你型计算机，而作为其核心的手机操作系统也成为人们讨论和研究的焦点。

1.1.1 Android 简要介绍

Android 本义指"机器人"，是基于 Linux 内核的手机软件平台和操作系统，是 Google 于 2007 年 11 月 5 日公布的手机系统平台，Android 早期由 Google 开发，后由开放手机联盟（Open Handset Alliance）开发，官方中文名为安卓。它采用了软件堆层（Software Stack，又称软件叠层）的架构；底层以 Linux 内核工作为基础，只提供基本功能，其他的应用软件则由各公司以 Java 语言作为编写程序的一部分自行开发。另外，为了推广此技术，Google 和其他几十个手机公司建立了开放手机联盟。Android 在未公开之前常被传闻为 Google 电话或 gPhone，大多数传闻认为 Google 开发的是自己的手机电话产品，而不是一套软件平台。到 2010 年 1 月，Google 才发行自家的品牌手机电话 Nexus One。由于 Android 系统的开源特性，很多制造商都在生产使用 Android 系统的设备，如三星、摩托罗拉、HTC、索爱、LG、小米、华为、魅族等。Android 系统除了运行在智能手机上之外，还可以用在平板电脑、电视、汽车、手表、眼镜等很多设备上。[1]

1.1.2 Android 发展历史

Google 于 2005 年并购了成立仅 22 个月的高科技企业 Android，展开了短信、手机检索、定位等业务，与此同时，基于 Linux 的通用平台也进入了开发阶段。2007 年 11 月，Google 与 84 家硬件制造商、软件开发商及电信营运商组建开放手机联盟共同研发改良 Android 系统。随后 Google 以 Apache 开源许可证的授权方式，发布了 Android 的源代码。

2008 年，Patrick Brady 于 Google I/O 大会上演讲"Anatomy & Physiology of an Android"，并提出了 Android HAL 架构图。HAL 以.so 文件的形式存在，可以把 Android framework 与 Linux kernel 分隔开。2011 年第一季度，Android 系统在全球的市场份额首次超过塞班系统，跃居全球第一[2]。目前，在微软 Windows Phone 还未成熟，苹果 iOS 平台不对外开放的情况下，Android 系统具有一定的发展优势和良好的发展前景。

1.1.3 Android 版本升级

Android 最早的一个版本 Android 1.0 Beta 发布于 2007 年 11 月 5 日，至今已经发布了多个更新。这些更新版本都在前一个版本的基础上修复了 bug 并且添加了前一个版本所没有的新功能。Android 系统通常都以食物命名，如 1.5 版叫作 Cupcake（纸杯蛋糕），1.6 版叫作 Donut（甜甜圈），Donut 把社交网络功能作为升级重点，在"手机的各种体验中"增加了社交网络元素。

2008 年 9 月 23 日，Android 系统中的第一个正式版本发布：Android 1.0（Astro，"铁臂阿童木"）。全球第一台 Android 设备 HTC Dream（G1）就搭载了 Android 1.0 系统。以下是 Android 1.0 系统所拥有的功能。

- **Android Market**：可以通过 Android Market 下载应用程序和获得程序更新。
- **网页浏览器**：可以完全还原并显示 HTML 和 XHTML 的网页，并且可以通过多点触控对网页进行放大、缩小。
- **照相机**：支持照相机和摄像头，但是这个版本没有选项用来改变照相机的分辨率、白平衡、质量等。
- 允许将应用程序图标放置到文件夹中，并且可以在主界面显示插件等东西。
- **支持 E-mail 传输**：支持 POP3、IMAP4 及 SMTP。
- **Gmail**：通过内置的 Gmail 应用程序进行 Gmail 同步。
- **Google 联系人**：通过 People 应用程序同步联系人。
- **Google 日历**：通过日历程序同步日历和日程。
- **Google 地图、Google 纵横及 Google 街景**：帮助用户查看地图和地理信息，并且可以通过 GPS 服务定位地理位置。
- **Google 同步**：一个管理 Android 设备中 Google 服务的应用功能。
- **Google 搜索**：允许用户在手机和网络上进行一致统一的搜索，包括联系人、电话、日历和信息等。
- **Google Talk**：一个聊天工具。
- 实时消息、语音信息和短信。
- **多媒体播放器**：负责管理、导入、复制和播放多媒体文件，但是不支持蓝牙耳机。
- 通知的信息可以在任务栏显示，并且可以对提示的方式进行设置，包括振动、声音、LED 或警告等。

- 声音识别器可以允许用户通过说话来输入文本、拨打电话，能更好地帮助残疾人士。
- 壁纸功能允许用户设置自己的照片和其他网络图片作为自己的手机主界面的背景。
- YouTube：内置 YouTube 在线应用程序。
- 其他应用程序：闹钟、计算器、电话、主界面、图库及设置。
- 支持 WiFi 和蓝牙。

2009 年 4 月 17 日，Google 正式推出 Android 1.5（Cupcake，"纸杯蛋糕"）。Android 1.5 最突出的功能非虚拟键盘莫属。当时，智能手机还主要依赖物理键盘进行输入，而这一点在 Android 1.5 中得到了改变。具体更新如下：

- 拍摄/播放视频，并支持上传到 Youtube；
- 支持立体声蓝牙耳机，同时改善自动配对性能；
- 采用 WebKit 技术的浏览器，支持复制/粘贴和页面中搜索；
- GPS 性能大大提高；
- 提供屏幕虚拟键盘；
- 主屏幕增加音乐播放器和相框 widgets；
- 应用程序自动随着手机旋转；
- 短信、Gmail、日历、浏览器的用户界面大幅改进，如 Gmail 可以批量删除邮件；
- 照相机启动速度加快，拍摄图片可以直接上传到 Picasa；
- 来电照片显示。

2009 年 9 月 15 日，Android 1.6（Donut，"甜甜圈"）SDK 发布，该版本基于 Linux 2.6.29 内核。其主要的更新如下：

- 重新设计的 Android Market；
- 手势支持；
- 支持 CDMA 网络；
- 文本转语音系统（Text-to-Speech）；
- 快速搜索框；
- 全新的拍照界面；
- 查看应用程序耗电；
- 支持虚拟私人网络（VPN）；
- 支持更多的屏幕分辨率；
- 支持 OpenCore2 媒体引擎；
- 新增面向视觉或听觉困难人群的易用性插件。

Android 的 6.0 版本，即 Android Marshmallow（简称 Android M）已经在 Google 2015 年的 I/O 大会上被正式发布。Android M 为工作升级而生（Android for Work Update）成为当时在业内被热议的话题。有业内人士解释道："Android M 将把 Android 的强大功能拓展至任何你所能看到的工作领域。"其主要新功能如下：

- 应用权限管理;
- SD 卡可以和内置存储"合并";
- Android Pay;
- 原生指纹识别认证;
- 自动应用数据备份;
- App Links(尽量减少诸如"你想要使用什么来打开这个?"的提醒);
- 打盹和应用待机功能;
- 可定制的 Quick Toggles 和其他 UI 调整;
- 可视化的语音邮件支持;
- 重新设计的时钟插件和音乐识别插件;
- 在设置中新出现的全新"Memory"选项条目(早期版本中出现过,不过后来被隐藏);
- 完成截图之后可以通过通知中心直接删除截图;
- Google Now Launcher 支持横屏模式;
- 带水平滚动条和垂直滚动条支持的全新应用和窗口小部件抽屉;
- 内置的文件管理器能够获得功能方面的明显升级;
- 支持原生点击唤醒功能;
- 可以选择"heads up"或"peeking"通知;
- 原生 4K 输出支持;
- 严格的 APK 安装文件验证;
- 支持 MIDI;
- USB Type-C 端口支持;
- 全新的启动动画;
- 引入"语音交互"API 在应用中提供更好的语音支持;
- 可通过语音命令切换到省电模式;
- 可以通过蓝牙键盘快捷方式来撤销和重做文本;
- 在联系人应用中能够对已经添加的联系人进行合并、删除或分享;
- 会有针对文本选择的浮动工具栏出现,以便于更快地选择文本;
- 默认应用的 UI;
- 允许通过分享菜单直接分享给联系人好友;
- 更细化的应用程序信息;
- 原生蓝牙手写笔支持;
- 分屏键盘;
- 移动的收音机;
- Mobile Radio Active 服务电池续航 bug 将会被修复;
- 除重复来电之外优化勿扰模式;

- 蓝牙扫描可用来改善定位精准度；
- 原生 Flashlight API；
- 更容易进行多种声音的设置（铃声、多媒体和闹钟）；
- 更平滑的声音滑块。

Android7.0 初次公开亮相于 2016 年 5 月 18 日的 Google I/O 大会。其主要新功能如下：
- 分屏多任务；
- 全新下拉快捷开关页；
- 通知消息快捷回复；
- 通知消息归拢；
- 夜间模式；
- 流量保护模式；
- 全新设置样式；
- 改进的 Doze 休眠机制；
- 系统级电话黑名单功能；
- 菜单键快速应用切换。[3]

2017 年，Google 发布了 Android 8.0 Oreo，包括了许多功能特性，如它设置了一个持续运行并消耗内存后台的服务。其新功能特性包括：
- 后台限制（Background Limit）；
- 通知频道（Notification Channel）；
- 自动填充 API（Auto-fill API）；
- 画中画（Picture-in-picture）；
- 自适应图标（Adaptive icons）；
- 字体（Fonts）和可下载字体（Downloadable Fonts）；
- 自动调整大小的 TextView（Auto-sizing TextView）。

1.2 Android 体系结构及特点

　　Android 平台采用了整合的策略思想，包括底层的 Linux 操作系统、中间层的中间件和上层的 Java 应用程序，Android 系统的体系结构如图 1-1 所示。Android 系统的特性及其体系结构总结如下。

　　（1）应用程序框架支持组件的重用与替换。便于用户把系统中不需要的应用程序删除，安装自己喜欢的应用程序。

　　（2）Dalvik 虚拟机专门为移动设备进行了优化。Android 应用程序将由 Java 编写、编译的类文件通过 DX 工具转换成一种后缀名为.dex 的文件来执行。Dalvik 虚拟机基于寄存器，相对于 Java 虚拟机速度要快很多。

（3）内部集成浏览器基于开源的 WebKit 引擎。有了内置的浏览器，意味着 WAP 应用的时代即将结束，真正的移动互联网时代已经来临，手机就是一台"小电脑"，可以在网上随意邀游。

（4）优化的图形库，包括 2D 和 3D 图形库，其中 3D 图形库基于 OpenGL ES 1.0。强大的图形库为游戏开发带来了福音。

（5）SQLite 用于结构化的数据存储。

（6）多媒体支持包括常见的音频、视频和静态映像文件格式，如 MPEG4、H.264、MP3、AAC、AMR、JGP、PNG、GIF。

（7）GSM 电话（依赖于硬件）。

（8）蓝牙（Bluetooth）、EDGE、3G、WiFi（依赖于硬件）。

（9）照相机、GPS、指南针和加速度计（依赖于硬件）。

（10）丰富的开发环境，包括设备模拟器、调试工具、内存及性能分析图表和 Eclipse 集成的开发环境插件。

Google 提供了 Android 开发包 SDK，其中包含大量的类库和开发工具，并且针对 Eclipse 的可视化开发了插件 ADT。

图 1-1 展示了 Android 系统的体系结构。由图可见，Android 系统的体系结构可分为 4 层，由上到下依次是应用程序、应用程序框架、核心类库和 Linux 内核，其中第三层还包括 Android 运行时的环境。

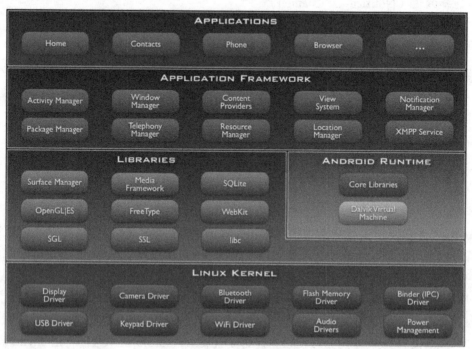

图 1-1 Android 系统的体系结构

1.2.1 应用程序

Java 为编程语言，使得 Android 从接口到功能都有层出不穷的变化。其中 Activity 等同于 J2ME 的 MIDlet，一个 Activity 类（Class）负责建立视窗（Window），Activity 通常在前台（Foreground）工作，而在后台（Background）运行的程序被称为 Service。Activity 和 Service 之间通过 ServiceConnection 和 AIDL 链接，达到复数程序同时运行的效果。如果运行中 Activity 的全部画面被其他 Activity 取代，则该 Activity 便被停止，甚至被系统清除（Kill）。

View 基本上等同于 J2ME 中的 Displayable，程序人员可以通过 View 类与"XML layout"文件来设置用户界面。从 Android 1.5 版本开始，可以利用 View 生成 Widget，Widget 是 View 的一种特例，因此可以使用 XML 文件来设计和配置 Widget 的 layout。例如，HTC 制造的 Android Hero 手机含有大量的 Widget。ViewGroup 则是各种 layout 的基础抽象类，ViewGroup 之内还可以嵌入 ViewGroup。通常，View 的构造函数不需要在 Activity 中调用（注意，在 Displayable 中是必需的）。在 Activity 中，一般要通过 findViewById()来从 XML 中获取。Android 系统中，View 与事件（Event）息息相关，两者之间通过 Listener 结合在一起，每一个 View 都可以注册一个 Event Listener。例如，当 View 要处理用户触碰（Touch）的事件时，就要向 Android 框架注册 View.OnClickListener。

1.2.2 中间件

操作系统与应用程序的沟通桥梁分为两层：函数层（Library）和虚拟机（Virtual Machine）。

函数层中的 Bionic 是 Android 改良 Libc 的版本。Android 同时包含了 Webkit，所谓的 Webkit 就是 Apple Safari 浏览器背后的引擎。Surface flinger 用于将 2D 或 3D 的内容显示到屏幕上。Android 使用的工具链（Toolchain）为 Google 自己研发的 Bionic Libc。

Android 采用 OpenCORE 作为基础多媒体框架。OpenCORE 可分为 7 大块：PVPlayer、PVAuthor、Codec、PacketVideo Multimedia Framework（PVMF）、Operating System Compatibility Library（OSCL）、Common、OpenMAX。

Android 使用 Skia 为内核图形引擎，搭配 OpenGL/ES。Skia 与 Linux Cairo 功能相当。2005 年，Skia 公司被 Google 收购；2007 年年初，Skia GL 源代码被公开，目前 Skia 也是 Google Chrome 的图形引擎。

Android 的多媒体数据库采用 SQLite3 数据库系统。数据库又分为共用数据库与私用数据库。用户可通过 ContentResolver 类（Column）获得共用数据库。

Android 的中间层多以 Java 实现，并且采用特殊的 Dalvik 虚拟机（Dalvik Virtual Machine）运行。Dalvik 虚拟机是一种"寄存器形态"（Register Based）的 Java 虚拟机，变量皆存放于寄存器中，虚拟机的指令相对减少。

Dalvik 虚拟机可以有多个 Instance，每个 Android 应用程序都由一个相对独立的 Dalvik 虚拟机来运行，可让系统在运行程序时达到优化。Dalvik 虚拟机并非运行 Java Bytecode，而是运行一种称为.dex 格式的文件。

1.2.3 硬件抽象层

Android 硬件抽象层 HAL（Hardware Abstract Layer）用于将 Android Framework 与 Linux Kernel 分隔开，降低对 Linux Kernel 的依赖，以实现 Kernel Independent。它目前以 HAL Stub 的形式存在，本身是.so 档，属于 proxy 的概念。Android Runtime 向 HAL 取得 Stub 的 Operations，再以 Callback 的方式操作函数。

1.2.4 操作系统

Android 运行于 Linux Kernel 之上，但并不是 GNU Linux。原因是 GNU Linux 里的功能 Android 大都没有支持，如 Cairo、X11、Alsa、FFmpeg、GTK、Pango、Glibc 等都被移除了。Android 又以 Bionic 取代了 Glibc，以 Skia 取代了 Cairo，再以 Opencore 取代了 FFmpeg 等。为了达到商业应用，Android 必须移除关于 GNU Copyleft 的限制，如 Android 将驱动程序移到 Userspace，使得 Linux Driver 与 Linux Kernel 彻底分开。

目前 Android 的 Linux Kernel 控制包括安全（Security）、存储器管理（Memory Management）、程序管理（Process Management）、网络堆栈（Network Stack）、驱动程序模型（Driver Model）等。[4]

1.3 其他主要智能手机开发平台

目前流行的智能手机开发平台有 iOS、Windows CE、Symbian OS、Palm 等。按照源代码、内核和应用环境等的开放程度划分，智能手机开发平台可分为开放型平台（基于 Linux 内核）和封闭型平台（基于 Windows 内核）两大类。

1.3.1 iOS 简介

iOS 是由苹果公司开发的移动操作系统，属于类 UNIX 的商业操作系统。iOSX 是有限的 SDK（软件开发工具包），保证程序员可以利用全功能 OS，而不必深入核心，也就是说，核心是不完全开放的。原本这个系统名为 iPhone OS，因为 iPad、iPhone、iPod touch 都使用 iPhone OS，所以 2010 年的 WWDC 大会上宣布将其改名为 iOS。苹果公司于 2014 年 WWDC（苹果开发者大会）发布的新开发语言 Swift，可与 Objective-C 共同运行于 Mac OS 和 iOS 平台，用于搭建基于苹果公司平台的应用程序。其语法内容混合了 OC、JS、Python，语法简单，使用方便。[5]

1.3.2　Windows CE 简介

　　Windows CE 虽然不是一个严格意义上的实时内核，却是专门为嵌入式系统设计的。它支持嵌套中断，允许更高优先级别的中断首先得到响应，而不需要等待低级别的 ISR（Interuption Service Routine，中断服务程序）完成。这使得该操作系统具有嵌入式操作系统所要求的实时性，同时有更好的线程响应能力。Windows CE 对高级别 IST（中断服务线程）的响应时间上限的要求更加严格，在线程响应能力方面进行了改进，可帮助开发人员掌握线程转换的具体时间，并通过增强的监控能力和对硬件的控制能力帮助他们创建新的嵌入式应用程序。另外，Windows CE 有 256 个优先级别，这样使得开发人员在控制嵌入式系统的时序安排方面有更大的灵活性。

　　Windows CE 是封闭的操作系统，其软件 Windows 是商业软件，它的源代码是企业的最高机密，因此不可能开放。使用类似于 Visual C++的软件，第三方可以开发应用 Windows CE。Windows CE 也利用了类似于视窗的.NET 框架，但所有的使用和服务都是收费的。

　　从硬件支持能力上看，Windows CE 支持 Arm、MIPS、X86 和 SuperH。

　　Windows CE 也已经发现了一些病毒感染的案例，病毒作者对微软平台的热衷在手机系统上得到了延续。Windows CE 很早就被发现存在安全漏洞，典型的漏洞是允许攻击者向使用该系统的手机发送恶意代码，这一点与基于 Windows 的 PC 系统非常相似。[5]

1.3.3　Symbian 简介

　　Symbian 9.0 及以后的版本使用了 EKA2 核心。EKA2 全面改进了原有的任务调度算法，完全支持实时性，支持某些高带宽、高优先级的任务对系统的基本实时性要求。这些任务包括 VoIP 网络电话、高速率的视频在线点播。EKA2 的改进有：内核实时（real-time）增强，多线程处理能力更好，API 调用更高效快速，使得 EKA2 成为一个真正意义上的 32 位操作系统。

　　Symbian 系统本身存在一些安全漏洞，因此其目前受病毒影响最深。已经发现的针对 Symbian 的病毒超过了 50 种，这些病毒通常感染 Symbian 6.0 系统，而 UIQ 平台极少发生感染。最为人所知的 Cabir 病毒是通过运用蓝牙连接对 Symbian 手机进行 DoS 攻击的。由于越来越多的个人信息（如电话簿、商业机密文档）会被保存在智能手机中，为了防止恶意软件或病毒窃取这些信息或耗费用户的通信费用，Symbian 9.0 及以后的版本引入了新的系统安全模型。Symbian 9.0 以前的系统中安装的某个软件的所有文件都会存储在\system\apps\xxx 目录下，Symbian 9.0 及以后的系统中安装的某个软件中的不同文件会存放在不同的目录下。例如，可执行文件（.exe，以前是.app）放在\sys\bin 下，资源文件放在\resource 下，每个软件的所有私有数据被放在\private\下，其他目录是供所有软件共享的目录。其中\sys\bin 和\resource 用户不能更改，可执行文件只能由安装程序复制进去。\private\只能由软件安全包所对应的软件访问。可执行文件引入了能力模型，取得某些能力如访问用户的电话簿、发送短信、修改手机设置等，需要让可执行文件获得 Symbian 公司或诺基亚公司的数字签名。该系统还具备可执行文件防篡改功能，安装经过修改的软件包，或者用读卡器修改存储卡的\sys\bin 目录，会被 Symbian 系统发现。引入这些特性，使得系统的安全性大大提高，各种私人数据可以放心地保存在手机中。但是安全模型的引入，却导致 Symbian 系统出现了兼容性问题。

Symbian 系统可以支持从 ARM9 系列到 ARM11 系列的所有 ARM 处理器。Symbian OS v9.5 是业界首款可支持 ARM Cortex-A8 处理器的智能手机操作系统。[5]

1.3.4　Palm OS 简介

Palm OS 是一套专门为掌上电脑开发的 OS。在编写程序时，Palm OS 充分考虑了掌上电脑内存相对较小的特性，因此它只占用非常小的内存。因为基于 Palm OS 编写的应用程序占用的空间也非常小（通常只有几十 KB），所以基于 Palm OS 的掌上电脑尽管只有几 MB 的 RAM，却可以运行众多应用程序，并且有较好的实时性能。同时，Palm OS 有着合理的内存管理，其存储器全部是可读写的快速 RAM。RAM 分为两种：动态 RAM 和静态 RAM。动态 RAM 类似于 PC 上的 RAM，它为全局变量和其他不需要永久保存的数据提供临时的存储空间；静态 RAM 类似于 PC 上的硬盘，可以永久保存应用程序和数据。

Palm OS 是一套具有极强开放性的系统。开发者免费向用户提供 Palm OS 的开发工具，允许用户利用该工具在 Palm OS 的基础上方便地编写、修改相关软件。[6]

Palm OS 支持的处理器有 Motorola DragonBall、Xscale 等。

第 2 章
Android 开发基础

2.1 面向对象编程介绍

面向对象编程（Object Oriented Programming, OOP）是一种计算机编程架构。OOP 的一条基本原则为计算机程序是由单个能够起到子程序作用的单元或对象组合而成的。OOP 达到了软件工程的三个主要目标：重用性、灵活性和扩展性。为了实现整体运算，每个对象都能够接收信息、处理数据和向其他对象发送信息。[7]

面向对象的三大基本特征是封装、继承、多态。下面逐一进行介绍。

封装：是一种信息隐蔽技术，由类的定义来体现，是对象的重要特性。封装使数据和加工该数据的方法成为一个整体，以实现独立性很强的模块，用户只能见到对象的外特性（对象能接收哪些消息，具有哪些处理能力），而对象的内特性（保存内部状态的私有数据和实现加工能力的算法）对用户是隐蔽的。封装的目的在于把对象的设计者和对象的使用者分开，使用者不必知晓行为实现的细节，只需用设计者提供的消息来访问该对象。[8]

继承：是子类自动共享父类之间数据和方法的机制。它由类的派生功能体现。一个类直接继承其他类的全部描述，同时可进行修改和扩充。继承具有传递性。继承分为单继承（一个子类只有一个父类）和多重继承（一个子类有多个父类）。类的对象是各自封闭的，如果没有继承性机制，则类对象中的数据、方法就会出现大量重复。继承不仅支持系统的可重用性，还促进系统的可扩充性。[9]

多态：对象根据所接收的消息做出动作，同一消息被不同的对象接收时可产生完全不同的行动，这种现象称为多态性。用户利用多态性可发送一个通用的信息，而将所有的实现细节都留给接收消息的对象自行决定，这样同一消息即可调用不同的方法。例如，Print 消息被发送给一张图或表时调用的打印方法与将同样的 Print 消息发送给一个正文文件时调用的打印方法完全不同。多态性的实现得到了继承性的支持，利用类继承的层次关系，把具有通用功能的协议存放在类层次中尽可能高的地方，而将实现这一功能的不同方法置于较低层次，这样，在这些低层次上生成的对象就能给通用消息以不同的响应。[10]

面向对象设计方法以对象为基础，利用特定的软件工具直接完成从对象客体的描述到软件结构之间的转换。这是面向对象设计方法最主要的特点和成就。面向对象设计方法的应用解决了传统结构化开发方法中客观世界描述工具与软件结构的不一致性问题,缩短了开发周期,解决了从分析和设计到软件模块结构之间多次转换映射的繁杂过程,是一种很有发展前途的系统开发方法。[11]

2.2 Android Java 基础

Android 系统的开发语言是 Java 语言,因此分析 Android 系统的开发,应先认识 Java 语言。Java 语言是一门面向对象编程语言,不仅吸收了 C++语言的各种优点,还摒弃了 C++语言难以理解的多继承、指针等概念,具有功能强大和简单易用两个特征。Java 语言作为静态面向对象编程语言的代表,极好地实现了面向对象理论,允许程序员以优雅的思维方式进行复杂的编程。[11]

Java 包含以下三种技术架构。

JAVAEE:Java Platform Enterprise Edition,开发企业环境下的应用程序,主要针对 Web 程序开发。

JAVASE:Java Platform Standard Edition,完成桌面应用程序的开发,是其他两者的基础。

JAVAME:Java Platform Micro Edition,开发电子消费产品和嵌入式设备,如手机中的程序。

作为计算机语言中发展得最快的语言,概括起来,Java 语言具有以下几个特点。

第一,Java 语言具有面向对象的特点,比较易于被人们所理解。现实中的任何实体都可以看作对象,并归属于某类事物中,也就是说,任何对象都是某类事物的事例。如果将传统的过程式编程语言解释为以过程为中心,以算法为驱动的一种"算法+数据"的程序编写语言,那么面向对象的 Java 语言可以说是以对象为中心,以消息为驱动的一种"对象+消息"的程序编写语言。此外,Java 语言具有很强的封装性。所谓封装,就是用一个自主式框架把对象的数据和方法组合成一个整体。面向对象的封装性、多态性和继承性,使得 Java 语言的交互功能越来越成熟。

第二,Java 语言具有可靠性和安全性的特点。Java 语言的最初设计目的是应用于电子类消费产品,因此其可靠性要求较高。Java 语言虽然源于 C++语言,但它克服了许多 C++语言的不可靠因素。例如,显式的方法声明可以确保编译器发现方法调用错误;不支持指针,可以防止内存的非法访问;自动单元收集可以避免内存丢失等导致的问题;解释器运行实时检查可以发现数组和字符串访问是否越界等,都保证了 Java 语言的可靠性。Java 语言通过自己的安全机制还可以防止病毒程序的产生,减轻下载程序对本地系统的威胁、破坏。在解码器内还有字节校检器进行检查,来自网络的类则由类装载器负责装载到单独的内存区,避免应用程序之间的相互干扰。这些机制使得 Java 语言成为安全的编程语言。

第三,Java 语言具有体系结构独立的特点。以往所通用的程序语言都有同一个弊端,即只能在统一体系的计算机结构中运行的弊端。而 Java 语言的运行与计算机结构无关,在任何系统中都可以随意运行。[11]

Java 语言的 Android 开发基本架构包括应用程序、应用程序框架、程序库、运行库和数据库。首先是应用程序。Android 手机软件会与同一系列的核心应用程序一起发布,它包括一些客户端,短消息程序,日历、地图、浏览器等管理程序。Java 语言作为 Android 程序的编写工具,大大提高了 Android 手机交互的可能性。应用程序框架是指开发人员访问核心应用程序所使用的主要框架。框架的设计减少了一些组件的重用频率,在遵循框架安全性原则的前提下,一个程序任何时候都可以发布可供任何其他应用程序使用的功能块。框架的重组机制还为用户替换程序组件提供了便利。Android 手机软件的数据库与其他系统的数据库相比有着极大的优点,更易于管理,更新方便快捷,因此使用 Java 语言的 Android 应用软件可以更好地满足用户需求。[12]

Android 虽然使用 Java 语言作为开发工具,但是在实际开发中,它还是与 Java SDK 有一些不同的地方。例如,一些 Java SDK 被 Android SDK 采用时,一小部分没有被用在 Android SDK 中,如界面部分。java.awt package 中只有 java.awt.font 被 Android SDK 所采用,其他的在 Android 平台的开发工程中不能使用。

2.3 XML 基础

XML 语言又称为可扩展标记语言,是一种用于标记电子文件使其具有结构性的标记语言。[13]

它可以用来标记数据、定义数据类型,是一种允许用户对自己的标记语言进行定义的源语言。它非常适合万维网传输,为描述和交换独立于应用程序或供应商的结构化数据提供了统一的方法。它是 Internet 环境中跨平台的、依赖于内容的技术,也是当今处理分布式结构信息的有效工具。

XML 语言的设计宗旨是传输数据,而不是显示数据。它的标签没有被预定义,需要用户自行定义标签。

XML 语言结合了标准通用标记语言和 HTML 语言的优点并消除了其缺点。它实现了标准通用标记语言的大部分功能,但比标准通用标记语言简单。XML 语言简单明了,能很方便地运用于部分开发程序中。它的简洁明了使其成为数据写入的共用语言,尽管不同的应用软件也支持其他的格式,但它们的共同之处就是都支持 XML 语言,也就是说,程序可以更方便地与 Windows、Mac OS、Linux 及其他平台下的信息交流,并以 XML 作为输入与输出格式。

以下例子是 Jani 写给 Tove 的便签,存储为 XML 格式:

```
<note>
<to>Tove</to>
```

```
<from>Jani</from>
<heading>Reminder</heading>
<body>Don't forget me this weekend!</body>
</note>
```

这条便签具有自我描述性，它包含了发送者和接收者的信息，同时拥有标题及消息主体。但是，此条便签实际上没有做任何事情，它只是以 XML 格式存储便签中的消息。想要上传、接收和显示此标签，使用者需要编写其他运行程序。

2.4　IDE Eclipse 介绍

集成开发环境（Integrated Development Environment，IDE）是用于提供程序开发环境的应用程序，一般包括代码编辑器、编译器、调试器和图形用户界面等工具。它是集代码编写功能、分析功能、编译功能、调试功能等于一体的开发软件服务套（组）。所有具备这一特性的软件或软件套（组）都可以叫作集成开发环境。[14]

Eclipse 是著名的跨平台开源集成开发环境。它最初主要用于 Java 语言的开发，目前也有人通过插件使其作为 C++、Python、PHP 等其他语言的开发工具。Eclipse 本身只是一个框架平台，但是众多插件的支持，使得 Eclipse 拥有较佳的灵活性，因此许多软件开发商以 Eclipse 为框架开发自己的 IDE。[14]

Eclipse 的插件机制是一种轻型软件组件化架构。在客户机平台上，Eclipse 使用插件来提供所有的附加功能，如支持 Java 以外的其他语言。已有的分离的插件已经能够支持 C/C++（CDT）、Perl、Ruby、Python、Telnet 和数据库开发。插件架构能够支持将任意扩展加入现有环境中，如配置管理，而不仅仅限于支持各种编程语言。

Eclipse 在设计上以插件为主。除核心组件并不大，其他所有的功能都以插件的形式实现，所以它的核心组件并不大。Eclipse 的基本内核包括：图形 API（SWT/Jface）、Java 开发环境插件（JDT）、插件开发环境（PDE）等。

Eclipse 是一个开放源代码的软件开发项目，专注于为高度集成的工具开发提供一个全功能的、具有商业品质的工业平台。它主要由 Eclipse 项目、Eclipse 工具项目和 Eclipse 技术项目组成，具体包括四部分：Eclipse Platform、JDT、CDT 和 PDE。JDT 支持 Java 开发、CDT 支持 C 开发、PDE 支持插件开发，Eclipse Platform 则是一个开放的可扩展 IDE，提供了一个通用的开发平台。Eclipse Platform 允许开发与其他工具无缝集成的工具。[15]

2.5　Android Studio 介绍

Android Studio 是一个集成开发工具，是基于 Intellij IDEA 的开发平台。与 Eclipse ADT

相似，Android Studio 提供了集成的 Android 开发工具以进行开发和调试。

在 IDEA 的基础上，Android Studio 在架构上提供了：

（1）基于 Gradle 的构建支持；

（2）Android 专属的重构和快速修复；

（3）具备提示工具，能够获取应用的性能、可用性、版本兼容性等；

（4）支持 ProGuard 和应用签名；

（5）基于模板的向导来生成常用的 Android 应用设计和组件；

（6）功能强大的布局编辑器，可以让用户拖曳 UI 控件并进行效果预览。[16]

2.6 Kotlin 介绍

Kotlin 是一个基于 JVM 的新的编程语言，由 JetBrains 开发，Android 8.0 已经开始支持它。Kotlin 既可以编译成 Java 字节码，也可以编译成 JavaScript 脚本，方便在没有 JVM 的设备上运行。JetBrains，作为目前广受欢迎的 Java IDE IntelliJ 的提供商，以 Apache 许可的方式对 Kotlin 编程语言进行了开源。Kotlin 已正式成为 Android 的官方支持开发语言。

相对 Java 来说，Kotlin 在编写代码时有如下优势：代码简洁高效、函数式编程、空指针安全、支持 lambda 表达式、流式 API 等。在执行效率上，Kotlin 和 Java 具有相同的理论速度（都是编译成 JVM 字节码）。另外，为了与存量项目代码和谐共处，Kotlin 和 Java 是互相完美兼容的，两种代码文件可以并存、代码可以互相调用、文件可以互相转换，库文件也可以无障碍地互相调用，也就是说，使用 Kotlin 基本不会带来额外的成本负担。

Kotlin 作为 Java 的改良语言，在 Android 开发中有很多优势。我们可以先从相对直观的界面绘制开始了解它，然后认识其语法特点，再慢慢去接触其更深层次的编程思想。

1. 简化 findViewById

在 Android 的架构里，XML 布局文件和 Activity 是松耦合的，Activity 中想要使用界面元素，必须借助 R 文件对 XML 控件的记录，用 findViewById 找到这个元素。在 Kotlin 中可以继续使用 findViewById 去绑定 XML 布局中的控件：（TextView）findViewById（R.id.hello）；进一步引用 Anko 之后，可以使用 find 函数去绑定控件：find（R.id.hello），不需要进行类型转换。

2. Anko

Anko 其实是一种 DSL（领域相关语言），专门用代码方式来定义界面和布局。这样就不需要使用 XML 布局文件。虽然在 XML 中定义界面布局有很多优点，如分层清晰、代码易读、容易预览效果等，但是它的渲染过程复杂，难以重用（虽然有 including），而如果我们用 Java 代码去替换 XML 布局文件，代码会更加复杂。而 Anko 实现了在代码中简洁优雅地定义界面和布局，而且由于不需要读取和解析 XML 布局文件，使得其性能表现更佳。

虽然 Anko 效率很高，代码简洁，清爽直观，但是目前还有很多缺点，主要包括：AS 并不支持直接预览 Anko 界面，虽然有一个 Anko DSL Preview 插件，并不能起到预览作用，需要 make 才能刷新，此外该插件和现在的 AS 不兼容，如果要在多版本中动态替换外部资源，需要使用动态类加载才能实现，无法借用资源 Apk 实现，不方便根据 View 的 id 去即时引用 View 控件。

第 3 章 搭建 Android 开发环境

3.1 Android SDK 介绍

3.1.1 Android SDK 目录结构

Android SDK 目录结构如图 3-1 所示。

其中，各个目录及文件功能如下。

- add-ons：包含 Google API，如 GoogleMaps。
- build-tools：包含各版本的 SDK 编译工具。
- docs：包含离线开发者文档、Android SDK API 参考文档。
- extras：包含扩展开发包，例如，可以使高版本的 API 在低版本的开发中使用。
- platforms：包含各版本 SDK。进入根据 API Level 划分的 SDK 版本（这里以 Android 2.2 为例）后有一个 android-8 的文件夹，进入 android-8 后是 Android 2.2 SDK 的主要文件，其中 ant 为 ant 编译脚本，data 中保存着一些系统资源，images 是模拟器映像文件，skins

图 3-1 Android SDK 目录结构

是 Android 模拟器的皮肤，templates 是工程创建的默认模板，android.jar 是该版本的主要 framework 文件，tools 目录里包含了重要的编译工具，如 aapt、aidl、逆向调试工具 dexdump 和编译脚本 dx。
- platform-tools：包含各版本 SDK 通用工具，如 adb、aapt、aidl、dx 等文件，Android 1、2、3 版本提示。这里和 platforms 目录中的 tools 文件夹有些重复，主要原因是从 Android 2.3 开始这些工具已被划分为通用的了。
- samples：包含各版本 API 使用样例。Android SDK 自带的默认示例工程，强烈推荐初学者运行样例进行学习。
- sources：包含各版本 SDK 源代码。

- **system-images**：包含模拟器映像文件。从 android-14 开始将模拟器映像文件整理在此处（原来放在 platforms 下）。
- **temp**：包含临时文件夹，一般在 SDK 更新安装时用到。
- **tools**：包含各版本 SDK 自带工具。例如，DDMS 用于启动 Android 调试工具，如 LoaCat、屏幕截图和文件管理器；draw9patch 是绘制 Android 平台的可缩放 png 图片的工具；sqlite3 可以在 PC 上操作 SQLite 数据库；monkeyrunner 是一个不错的压力测试应用，模拟用户随机按键；mksdcard 是模拟器 SD 卡映像的创建工具；emulator 是 Android SDK 模拟器主程序，不过从 Android 1.5 开始，需要输入合适的参数才能启动模拟器；traceview 是 Android 平台上重要的调试工具。
- **AVD Manager.exe**：Android 手机的模拟配置工具，用于配置模拟器，在此需要注意，只有先配置 AVD 才可运行模拟器。
- **SDK Manager.exe**：SDK 管理器，用于 SDK 的更新、下载、删除。

3.1.2 android.jar 内部结构

在 platforms 目录下的每个 Android 版本中，都有一个名为 android.jar 的文件。android.jar 是一个标准的压缩包，里面包含了编译后的压缩文件和全部的 API。使用解压缩工具可以打开此压缩文件，解压后可以看到其内部结构。android.jar 内部结构如表 3-1 所示。

表 3-1 android.jar 内部结构

android	包含用于标准 Android 开发的所有类
android.accessibilityservice	是一个辅助工具类，可以获取手机当前页面的信息、用户的操作事件等
android.accounts	Android 2.0 中加入的一个新包，该包主要包括了集中式的账户管理 API，用以安全地存储和访问认证的令牌和密码
android.app	高层模型，用于所有的 Android 程序
android.appwidget	UI 类，很多应用都要用到它，但属于高层方面
android.bluetooth	提供蓝牙服务
android.content	在设备上访问和发布数据
android.content.pm	访问一个应用程序包的信息，包括有关的活动、权限、服务、签名和提供者信息的类
android.content.res	访问文件，如原始资源、颜色、drawables、媒体或其他包中的其他文件，以及重要设备的配置细节（应用资源、输入类型等）
android.database	包含访问数据库的类
android.database.sqlite	包含 SQLite 数据库管理的类
android.gesture	提供类来创建、识别、加载和保存手势
android.graphics	提供在屏幕上绘图的底层工具
android.graphics.drawable	提供类来管理绘图元素
android.graphics.drawable.shapes	包含用于绘制几何图形的类
android.hardware	提供对不常用硬件的支持
android.inputmethodservice	文字输入法的基本类
android.location	定义本地化服务的类

续表

android.media	提供管理和播放视频或音频的类
android.net	在 java.net.*的基础上提供帮助联网的类
android.net.http	Android SDK 中一些与网络 HTTP 有关的 package
android.net.wifi	提供管理 WiFi 设备的类
android.opengl	提供 OpenGL 工具
android.os	提供基本的系统服务
android.preference	提供管理类、应用程序首选项和执行设置的用户界面
android.provider	提供方便的类来实现 Android 提供的服务
android.sax	一个可以方便地编写高效和强劲的 SAX 处理程序的框架
android.service.wallpaper	Android 中壁纸开发用到的相关服务包
android.speech	语音识别相关的 package
android.speech.tts	文本转语音相关的 package
android.telephony	提供 API 来检测基本的通信功能
android.telephony.cdma	提供 API 来实现 CDMA 的管理
android.telephony.gsm	提供 API 来实现 GSM 的管理
android.test	Android 测试框架

3.1.3 android.bat 批处理常用命令

1. 全局相关

- v - verbose：详细显示错误、警告和非正式信息。
- h - help：显示帮助文档。
- s - silent：安静模式，只有发生错误时才显示信息。

2. 正确的命令格式及含义

- list：列出存在的级别和 AVD。
- list avd：列出现有的 AVD。
- list target：列出可选的级别。
- create avd：创建新的 AVD。
- move avd：移动或重命名 AVD。
- delete avd：删除 AVD。
- update avd：升级 AVD 以匹配新的文件夹。

3. 创建一个工程

- create project：新建一个 Android 项目。
- update project：升级一个 Android 项目。
- create test-project：新建一个 Android 测试项目。
- update test-project：升级一个 Android 测试项目。
- update adb：升级 ADB 以支持定义在 SDK add-ons 里的 USB 设备。

- update sdk：升级 SDK 到新的平台。

3.1.4 模拟 SD 卡

在 Andorid 开发中经常遇到与 SD 卡有关的调试，如播放 MP3 文件、图片文件等，模拟 SD 卡需要建立 SD 卡镜像文件。在使用模拟器开发时，可以通过硬盘来模拟 SD 卡。Android 模拟 SD 卡的具体做法如下。

1．创建一个 SD 卡镜像文件

打开 cmd，进入 c 盘根目录下，输入如下命令：mksdcard 1024M sdcard.img。

命令格式：mksdcard <size>[<file>]。

参数说明如下：

size：要创建 SD 卡的大小，可以用 K 或 M 单位来表示。

file：SD 卡镜像的路径。

1024M 表示 1024 兆，即该 SD 卡有 1G 的容量；也可用 K 做单位（1M=1024K），K 和 M 必须大写。Android 支持 8M~128G 的 SD 卡。[17]

2．关联 SD 卡和模拟器

将 SD 卡和模拟器联系起来的就是所对应 SD 卡的路径。在 Android Studio 平台上，单击 "Tools→android→Android AVD Manager"，弹出一个窗口，在其中的 SD 卡选项中选择 file，在后面的框中输入 "c:\sdcard.img"，即第一步创建的 SD 卡镜像文件的位置，然后单击 create avd 即可。这样就将 SD 卡和模拟器进行了关联。

3．向 SD 卡中导入文件

这一步需要运行模拟器。在 cmd 下输入命令 "adb push test.mp3 sdcard/test.mp3"，该命令会将本地当前目录下的 test.mp3 文件复制到 sdcard 中，文件名不变。前面一个 test.mp3 是本地文件所在的路径，sdcard 是目的 SD 卡镜像的文件名。

4．在模拟器中使用 SD 卡中的文件

导入文件之后，如果想要在模拟器中访问此文件，还需要在模拟器的 Dev tools 中用 Media scanner 扫描一下媒体文件。

3.1.5 Traceview 工具

Traceview 是 Android 平台装备的一个很便捷的性能分析工具。它可以通过图形化的方式展示需要跟踪的程序的性能，能精确到 method。

首先，需要在程序中写入代码，以便于生成 trace 文件。利用 trace 文件可以将需要跟踪的程序转化为图形的形式。例如，在 activity 的 onCreate()中添加 Debug.startMethodTracing()，在 onDestroy()中添加 Debug.stopMethodTracing()，如图 3-2 所示。

```
79        /** Called when the activity is first created. */
80Θ       @Override
81        public void onCreate(Bundle savedInstanceState) {
82            super.onCreate(savedInstanceState);
83            setContentView(R.layout.main);
84
85            Debug.startMethodTracing(); //在此开始调试
86        }
87
88Θ       @Override
89        protected void onDestroy() {
90
91            Debug.stopMethodTracing(); //在此结束调试
92        }
```

图 3-2　trace 文件

然后，需要创建一个带有 SD 卡的 AVD，这样才能将 trace 文件保存到/sdcard/...中。既可以在命令中分别单独创建，也可以在创建 AVD 时一起将 SD 卡创建出来。创建之后通过 DDMS file explore 就可以看到/sdcard/目录下有一个 trace 文件，如果没有在 Debug 语句中设置名字则默认为 dmtrace.trace。现在把这个文件复制到计算机上指定的目录下。[18] 假设是 d:\目录下：

```
C:\Users\Administrator>adb pull /sdcard/dmtrace.trace d:\
418 KB/s (84865 bytes in 0.198s)
```

最后，可以通过命令行来执行 traceview。具体步骤为进入 tools 目录后，执行 traceview：

```
C:\Users\Administrator>traceview d:\dmtrace.trace
```

3.1.6　ADB 工具

ADB 的全名是 Andorid Debug Bridge，顾名思义，这是一个 Debug 工具，用于调试手机。它包含如下几部分。

- Client 端：运行在开发机器中，用来发送 ADB 命令。
- Deamon 守护进程：运行在调试设备中，调试手机或模拟器。
- Server 端：作为一个后台进程运行在开发机器中，用来管理 PC 中的 Client 端和手机的 Deamon 之间的通信。

常用的命令选项如下。

-d：将命令指向唯一连接的 USB 设备，如果存在多个 USB 设备，则返回错误。

-e：将命令指向唯一运行的模拟器，如果有多个模拟器正在运行，则返回错误。

-s <序列号>：通过指定序列号将命令指向 USB 设备或仿真器。

-p<产品名称或路径>：简单的产品名称，如'sooner'，或者是产品出口目录的相对/绝对路径，如'out/target/product/ sooner'。如果没有指定-p，则使用 ANDROID_PRODUCT_ OUT 设置环境变量，该变量需要的是绝对路径。

3.2 搭建开发环境

Android Studio 是当前 Android 的主要应用开发环境。本节主要介绍 Android Studio 的安装步骤。

3.2.1 安装 JDK

Android Studio 的运行需要 JRE 的支持。在 Windows 上安装 JRE/JDK 非常简单，首先需要在 Sun 官方网站下载，网址为 http://developers.sun.com/downloads/。在 JDK 的安装过程中会出现两次安装提示，分别是安装 JDK 及安装 JRE。它们都需要安装在同一个 Java 文件夹下的不同文件夹中（不能同时安装在 Java 文件夹的根目录下，JDK 和 JRE 安装在同一文件夹中会出错）。确定好安装路径后，一直单击"下一步"按钮，出现"finish"即完成安装。

安装完 JDK 后需要配置环境变量：
（1）计算机→属性→高级系统设置→高级→环境变量。
（2）系统变量→新建 JAVA_HOME 变量，变量值填写 JDK 的安装目录。
（3）系统变量→Path 变量→编辑，在变量值最后输入：%JAVA_HOME%\bin;%JAVA_HOME%\jre\bin;（注意原来 Path 的变量值末尾有没有"；"，如果没有需要先输入"；"再输入上面的代码）。
（4）系统变量→新建 CLASSPATH 变量，变量值填写：.;%JAVA_HOME%\lib;%JAVA_HOME%\lib\tools.jar（注意最前面有一点）。
（5）检验是否配置成运行 cmd 模式，输入：java -version（java 和-version 之间有空格），若如图 3-3 所示显示版本信息，则说明安装和配置成功。[19]

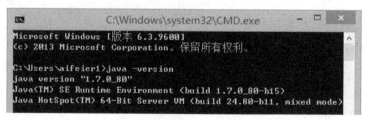

图 3-3　JDK 环境变量配置成功

3.2.2 安装 Android Studio

首先在 Android 官网 http://www.android-studio.org/index.php/download/hisversion 上下载安装包，然后双击安装包进行安装，如图 3-4 所示。

单击"Next"按钮，勾选"Android SDK"和"Android Virtual Device"，如图 3-5 所示。

图 3-4　安装 Android Studio（1）

单击"Next"按钮，设置 Android Studio 和 Android SDK 的安装路径（注意两者不要安装在同一路径下），如图 3-6 所示。

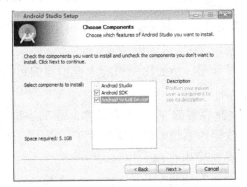

图 3-5　安装 Android Studio（2）

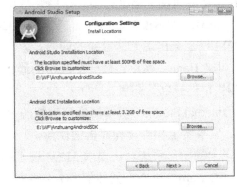

图 3-6　安装 Android Studio（3）

单击"Next"按钮，再选择"Install"就开始安装，如图 3-7 和图 3-8 所示。

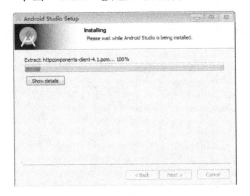

图 3-7　安装 Android Studio（4）

图 3-8　安装 Android Studio（5）

3.2.3　创建 Android 虚拟设备

AVD 的全称为 Android Virtual Device，就是 Android 的内部虚拟设备。它是 Android 的模拟器，可以运行建立的 Android 工程。每个 AVD 上可以配置多个需要运行的项目。

接下来介绍创建 AVD 的具体过程：
（1）在 Android Studio 窗口中单击 "Tools→Android→AVD Manager"。
（2）在弹出的窗口中单击 "Create Virtual Device"。
（3）接下来进行 AVD 的功能选择。

AVD name：要填写的虚拟机名称，由纯英文和数字组成。

Device：选择适合的屏幕大小、分辨率，如可以选择 3.2*QVGA（ADP2）（320*480:mdpi）。

Target：用来选择模拟器版本。例如，可以选择 Android4.0.3API15。

CPI/Abi：模拟器模拟的 CPU 型号，这里选 ARM（ARM 是手机的 CPU 型号）。

Keyboard 和 Skin：可以默认勾选，表示模拟器的界面上显示什么元素，这个不影响模拟器的使用，因为 Android 是触屏的。

Front camera 和 Back camera：表示是否模拟前置和后置摄像头，这里建议不选，有时选择该功能后模拟器会打不开，因此默认为 None 即可。

RAM：表示模拟器内存，512M、1024M 或者更高，但是 Google 要求不超过 1024M，1024M 以上会出现无法启动的问题。

VM Heap：模拟器每一个应用的最大内存空间分配，一般这个值越大模拟器的运行速度越快，选择默认即可。

Internal Storage：表示模拟器模拟的手机芯片存储容量的大小，就是通常所说的手机"内存"，它事实上相当于计算机的硬盘，一般用来存放操作系统，建议选择大于 1024M，因为 4.0 以上的操作系统默认要求 512M 以上内存，因此这个内存越大越好。

SD Card：用来设置外部 SD 卡的容量大小，会在默认目录里自动建立一个模拟 SD 卡的文件。

Snapshot：表示模拟器是否使用截图启动，这个功能可能会造成模拟器无法启动，但会使勾选上模拟器的启动加快。建议不勾选。

Use Host GPU：表示是否使用 PC 的 GPU 模拟手机显卡。建议不勾选，有些显卡模拟后无法启动模拟器。

设置好各个选项后，单击"确定"按钮即可。

3.3 DDMS 工具

随着 Android Studio 的广泛使用，开发人员对相关工具的使用需求更加凸显。DDMS（Dalvik Debug Monitor Service）是 Android 开发环境中的 Dalvik 虚拟机调试监控服务。它可以进行的操作有：为测试设备截屏、查看特定行程中正在运行的线程及堆信息、LogCat、广播状态信息、模拟电话呼叫、接收 SMS、虚拟地理坐标等。DDMS 功能非常强大，对于 Android 开发者来说是一个非常好的工具。[20]

单击 "Tools→Android→Android Device Monitor"，可以打开 DDMS 工具，如图 3-9 所示。

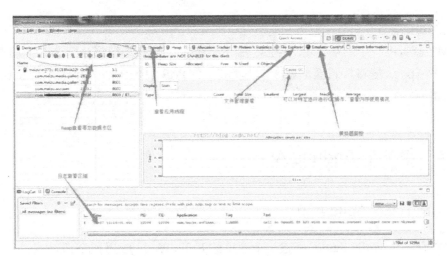

图 3-9　打开 DDMS 工具

3.3.1　DDMS 详细功能

（1）Devices：查看所有与 DDMS 连接的模拟器的详细信息，以及每个模拟器正在运行的 App 进程，每个进程最右边相对应的是与调试器连接的端口。

（2）Emulator Control：实现对模拟器的控制，如接听电话，根据选项模拟各种不同网络情况，模拟短信发送及虚拟地址坐标用于测试 GPS 功能等。

（3）LogCat：用来查看日志信息，可以对日志信息进行分类以方便查看。

（4）File Exporler：文件浏览器，用于查看 Android 模拟器中的文件，可以很方便地导入/导出文件。

（5）Heap：查看应用中的内存使用情况。

（6）Dump HPROF file：单击 DDMS 工具条上的"Dump HPROF"文件按钮，选择文件存储位置，然后运行"hprof-conv"。可以用 MAT 分析 heap dumps。启动 MAT，然后加载刚才生成的 HPROF 文件。MAT 是一个强大的工具，如有一种可以用来检测泄露的方法：直方图（Histogram）视图。它显示了一个可以排序的类实例的列表，内容包括：shallow heap（所有实例的内存使用总和），或者 retained heap（所有类实例被分配的内存总和，里面也包括它们所有引用的对象）等。

（7）Screen captrue：截屏操作。

（8）Thread：查看进程中的线程情况。[21]

3.3.2　DDMS 工作原理

DDMS 主要用于将 IDE 与测试终端（模拟器或链接设备）连接起来，它们应用各自独立的端口监听调试器的信息，DDMS 可以实时监测到测试终端的连接情况。当有新的测试终端连接后，DDMS 将捕捉到终端的 id，并通过 ADB 建立调试器，从而实现发送指令到测试终端的目的。

DDMS 监听第一个终端 App 进程的端口为 8600，APP 进程将分配 8601，如果有更多终端或更多 App 进程将按照这个顺序依次类推。DDMS 通过 8700 端口（base port）接收所有终端的指令。[22]

3.4 第一个 Android App

3.4.1 创建 Hello World App

下面将使用 Android Studio 创建一个简单的 Hello World 应用程序。

（1）打开 Android Studio，加载界面如图 3-10 所示。

（2）选择"Start a new Android Studio project"，如图 3-11 所示。

图 3-10 加载界面

图 3-11 创建 Android Studio 工程

（3）输入应用程序名、选择项目路径，然后单击"Next"按钮，如图 3-12 所示。

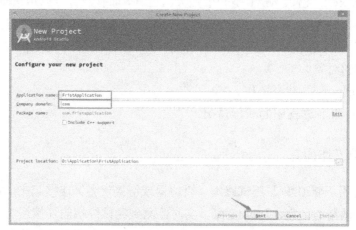

图 3-12 设置工程名称

(4)选择最低版本的 SDK，然后单击"Next"按钮，如图 3-13 所示。

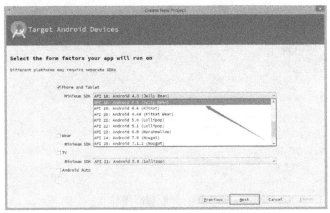

图 3-13　设置最低版本

(5)选择"Basic Activity"，然后单击"Next"按钮，如图 3-14 所示。

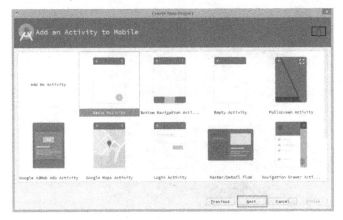

图 3-14　选择"Basic Activity"

(6)输入 Activity 名称、布局名称、标题等信息后，单击"Finish"按钮，如图 3-15 所示。

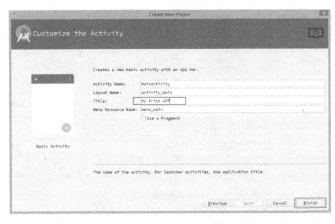

图 3-15　输入 Activity 名称、布局名称和标题

（7）在创建过程中加载相应组件，如图 3-16 所示。

图 3-16　加载相应组件

（8）系统窗口视图显示如图 3-17 所示。

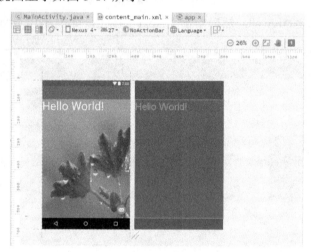

图 3-17　系统窗口视图显示

（9）布局 XML，设置背景、文字颜色和字体等，如图 3-18 所示。

图 3-18　布局 XML

（10）单击"运行"图标，选择已连接的运行设备，单击"OK"按钮，如图 3-19 所示。

（11）运行结果如图 3-20 所示。

图 3-19　运行程序

图 3-20　运行结果

3.4.2　Android 工程目录结构

Android 工程目录结构如图 3-21 所示。

图 3-21　Android 工程目录结构

-- src：源代码。如果最初选择创建 Activity，会有一个 Activity 的子类。

-- libs：工程中使用的库。

-- res：系统资源，所有文件都会在 R 文件生成资源 ID。

-- drawable：图片。

-- layout：界面布局。

-- menu：菜单。

-- mipmap-hdpi：高分辨率的图片目录。

-- mipmap -mdpi：中分辨率的图片目录。

-- mipmap -xhdpi：大分辨率的图片目录。

-- mipmap -xxhdpi：超大分辨率的图片目录。

-- mipmap -xxxhdpi：超超大分辨率的图片目录。

-- values：字符串、样式等数据。

-- AndroidManifest.xml：应用清单文件 Android 中的四大组件（Activity、ContentProvider、BroadcastReceiver、Service）都需要在该文件中注册。程序所需的权限也需要在该文件中声明，如电话、短信、互联网、SD 卡。

3.4.3 Android 程序部署与启动

1．应用程序安装、发布

在 Package Explorer 中用鼠标右键单击"工程→Run As→Android Application"即可完成程序的安装、发布，其快捷键为 Ctrl+F11。

2．程序启动过程

- 将.java 源文件编译成.class。
- 用 dx 工具将所有.class 文件转换为.dex 文件。
- 再将.dex 文件和所有资源打包成.apk 文件。
- 将.apk 文件上传并安装到模拟器中，存储在/data/app 目录下。
- 启动程序，开启进程。
- 根据 AndroidManifest.xml 文件找到 MainActivity 类，创建 Activity。
- Activity 创建后执行 onCreate（Bundle）方法，根据 R.layout.activity_main 构建界面。
- R.layout.activity_main 是 R 类中的一个成员变量，指向 res/layout/activity_main.xml 文件。
- activity_main.xml 文件中描述了 Activity 的布局方式及界面组件。
- 解析 activity_main.xml，通过反射创建对象，生成界面。

3.4.4 Android 程序打包安装过程

Android 程序打包安装过程如图 3-22 所示。

步骤：Android 程序→编译打包→apk→签名→通过 ADB 发布到设备上。

编译打包：dx.bat；可以将 bin 目录下的 classes 文件、deseLibs 依赖包打包成.dex 文件。还可将.dex 文件、工程的资源文件和清单文件打包成.apk 和签名文件（META-INF）（apk 实际上是一个.zip 文件）。

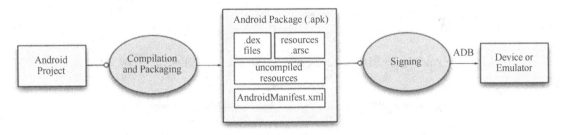

图 3-22　Android 程序打包安装过程

3.5　NDK 开发工具

　　Android 平台从诞生起，就已经支持 C、C++的开发。Android 的 SDK 基于 Java 实现，这意味着基于 Android SDK 进行开发的第三方应用都必须使用 Java 语言。但这并不等同于"第三方应用只能使用 Java 语言"。在 Android SDK 首次发布时，Google 就宣称其虚拟机 Dalvik 支持 JNI 编程方式，也就是第三方应用完全可以通过 JNI 调用自己的 C 动态库，即在 Android 平台上，"Java+C"的编程方式是一直都可以实现的。

　　NDK 的全称是 Native Development Kit。NDK 的发布，使得"Java+C"成为官方支持的开发方式。NDK 将是 Android 平台支持 C 开发的开端。使用 NDK 开发的优势包括以下几个方面。

　　（1）代码的安全性方面。apk 的 Java 层代码很容易被反编译，而 C/C++ 库的反编译难度较大。

　　（2）可以方便地使用现存的开源库。大部分现存的开源库都是用 C/C++代码编写的。

　　（3）能够提高程序的执行效率。将要求高性能的应用逻辑使用 C 开发，可提高应用程序的执行效率。

　　（4）便于移植。用 C/C++编写的库可以方便地在其他嵌入式平台上再次使用。

　　NDK 是一系列工具的集合，可帮助开发者快速开发 C（或 C++）的动态库，并能自动将 so 和 Java 应用一起打包成 apk。这些工具对开发者的帮助是巨大的。NDK 集成了交叉编译器，并提供了相应的 mk 文件来屏蔽 CPU、平台、ABI 等引起的差异，开发人员只需要简单修改 mk 文件（指出"哪些文件需要编译"、"编译特性要求"等），就可以创建出 so。NDK 可以自动将 so 和 Java 应用一起打包，极大地减轻了开发人员的打包工作。

　　但是，相比 Dalvik 虚拟机使用原生 SDK 编程也有一些劣势，如在 Android SDK 文档里找不到任何 JNI 方面的帮助。即使第三方应用开发者使用 JNI 完成了自己的 C 动态链接库（so）开发，但是 so 如何和应用程序一起打包成 apk 并发布存在一定的技术障碍，如程序更加复杂，兼容性难以保障，无法访问 Framework API，Debug 难度更大等。开发者需要自行斟酌使用。[23]

3.5.1 NDK 下载

单击"Tools→Android→SDK Manager→SDK Tools→NDK",下载 NDK,如图 3-23 所示。

图 3-23 NDK 的下载

3.5.2 NDK 开发

本书将在第 12 章详细讲解 NDK 开发的步骤和例程。

第 2 篇

Android 基础编程

- 第 4 章　Android App 基本概念
- 第 5 章　Android 应用用户界面设计
- 第 6 章　Android 图形编程
- 第 7 章　Android 数据存储编程

第 4 章

Android App 基本概念

Android 是用于移动设备的软件栈，包括操作系统、中间件和关键应用程序。Android SDK 为 Java 编程语言在 Android 平台上开发应用程序所需的工具和 API。

4.1　Android 应用中的基本概念

一个 Android 程序由以下部分组成：Activity、View、Service、Content Provider、Intent、Broadcast 等。Activity 代表 Android 程序的展现层，如用户看到的界面。一个 Android 程序会有多个 Activity，在程序运行过程中也会切换。View 是关于用户界面的视图布局，Activitiy 的用户界面通常继承自 android.view.View，View 的布局通过 android.view.ViewGroups 来管理。Service 是不需要 UI 展现的后台任务，可以通过 Android 系统的提醒框架给用户提示。Content Provider 用来为程序提供数据，通过 Content Provider 程序可以与其他程序共享数据。Android 系统的 SQLite 数据库可以看成一个 Content Provider。Intent 是一个异步的消息系统，可以从其他系统或服务获取数据，程序可以直接调用一个服务或 Activity，也可以请求 Android 系统服务。Broadcast Receiver 用来接收系统消息或隐含的 Intent，可以根据系统的改变做出反应。一个程序可以注册成为某些事件的 Broadcast Receiver，当事件发生时，程序就执行。

4.1.1　Activity

Activity 通常是应用程序中的单个屏幕，每个 Activity 都是作为扩展 Activity 基类的单个类来实现的。移动转换到另一个屏幕界面是通过开始一个新的 Activity 来实现的。在某些情况下，一个 Activity 可能会返回一个值以回到前一个 Activity 界面。当一个新的屏幕打开时，前一个屏幕会暂停并推送到历史堆栈里。Android 系统保留了从主屏幕启动的每个应用程序的历史堆栈，因此用户可以通过历史打开的屏幕向后导航。如果不需要保留，也可以选择从历史堆栈中删除界面。

接下来介绍 Activity 的属性。

1．android：alwaysRetainTaskState

是否保留状态不变。例如，切换回 Home，再重新打开，则 Activity 处于最后的状态。当一个浏览器拥有很多状态（当打开了多个 TAB 时），用户并不希望丢失这些状态时，可将此属性设置为 true。

2．android：clearTaskOnLaunch

用来表明当再次启动 Task 时，系统是否会清理除了根 Activity 以外的所有 Activity。默认值是 False。

3．android：configChanges

可以在 manifest.xml 文件中指定 Android 参数 configChanges，用于捕获手机状态的改变，如屏幕方向（orientation）的变化。在 Activity 中添加了 android:configChanges 属性后，当所指定属性发生改变时，会通知程序调用 onConfigurationChanged()函数。设置方法是将字段用"|"符号分隔开，如"locale|navigation|orientation"。在 Android 默认情况下，当 orientation 变化时会销毁当前 Activity，然后创建新的 Activity。如果不希望重新创建 Activity，可以在 AndroidManifest.xml 中配置 Activity：<activity android:name=".MainActivity" android:config Changes="|orientation" >，这样就不会在 orientation 变化时销毁重建 Acivity，而是调用 Activity 的 onConfigurationChanged()方法。

4．android：excludeFromRecents

是否可被显示在最近打开的 Activity 列表里。默认值是 False。

5．android：finishOnTaskLaunch

当用户重新启动这个任务时，是否关闭已打开的 Activity。默认值是 False。

这个属性与上面的 clearTaskOnLaunch 很像，不过它是指单个 Activity，而不是整个栈。当设置为 True 时，Task 重启后，这个 Activity 就不显示了。默认值是 False。

6．android：launchMode（Activity 加载模式）

在多 Activity 开发中，有可能是自己应用之间的 Activity 跳转，或者夹带其他应用的可复用 Activity。如果希望跳转到原来某个 Activity 实例，而不是产生大量重复的 Activity。这就需要为 Activity 配置特定的加载模式，而不是使用默认的加载模式。[24]

Activity 有四种加载模式：standard、singleTop、singleTask、singleInstance（其中前两个是一组、后两个是一组），默认为 standard。

standard：这是默认模式，可以不用写配置。在这个模式下，都会默认创建一个新的实例。因此，在这种模式下，每次跳转都会生成新的 Activity。此模式下可以有多个相同的实例，也允许多个相同的 Activity 叠加。

singleTop：也是发送新的实例，但 singleTop 不同于 standard 的一点是，当请求的 Activity 正好位于栈顶时（配置成 singleTop 的 Activity），不会构造新的实例。

singleTask：和后面的 singleInstance 为一组，都只创建一个实例，当 Intent 到来，需要创建设置为 singleTask 的 Activity 时，系统会检查栈里是否已经有该 Activity 的实例，如果有则直接将 Intent 发送给它。

singleInstance：首先说明一下 Task 这个概念，Task 可以被认为是一个栈，可放入多个 Activity。例如，如果要启动一个应用，Android 会创建一个 Task，随后这个应用的入口即被启动，如果调用其他的 Activity 也在同一个 Task 里面。如果在多个 Task 中共享一个 Activity 该如何处理呢？举个例子，如果开启一个导游服务类的应用程序，里面有一个 Activity 是开启 Google 地图的，当按下 Home 键退回到主菜单又启动 Google 地图的应用时，显示的就是刚才的地图，这实际上是因为它们是同一个 Activity，即引入了 singleInstance。singleInstance 模式就是将该 Activity 单独放入一个栈中，这个栈中只有这一个 Activity，不同应用的 Intent 都由这个 Activity 接收和展示，这样就做到了共享。当然前提是这些应用都没有被销毁，按下 Home 键即保存了这些应用，如果按下了返回键，则无效。[25]

7. android：multiprocess

是否允许多进程。默认值是 False。

8. android：noHistory

当用户从 Activity 上离开并且它在屏幕上不再可见时，Activity 是否从 Activity stack 中清除并结束。默认值是 False，Activity 不会留下历史痕迹。

9. android：screenOrientation

Activity 显示的模式。默认值是 unspecified。由系统自动判断显示方向，landscape 为横屏模式，宽度比高度大；portrait 为竖屏模式，高度比宽度大。user 模式：用户当前首选的方向。beind 模式：和该 Activity 下面的那个 Activity 的方向一致（在 Activity 堆栈中）。ensor 模式：由物理的感应器来决定，如果用户旋转设备则屏幕会横竖屏切换。nosensor 模式：忽略物理感应器，这样就不会随着用户旋转设备而更改了。

10. android：stateNotNeeded

Activity 被销毁或成功重启时是否保存状态。

11. android：windowSoftInputMode

Activity 主窗口与软键盘的交互模式，可以用来避免输入法面板遮挡问题。[24]

通常情况下，Activity 有三种状态：当它在屏幕前台时，响应用户操作的 Activity，处于激活或运行状态；当它上面有另外一个 Activity，使它失去了焦点但仍然对用户可见时，处于暂停状态；当它完全被另一个 Activity 覆盖则处于停止状态。Activity 从一种状态转变到另一种状态时，会调用其生命周期方法。Activity 一共有 7 种生命周期方法，具体如表 4-1 所示。

表 4-1 Activity 生命周期方法

方 法 名	说 明
void onCreate()	设置布局及进行初始化操作
void onStart()	可见，但不可交互
void onRestart()	调用 onStart()
void onResume()	可见，可交互
void onPause()	部分可见，不可交互
void onStop()	完全不可见
void onDestroy()	销毁

- startActivity 开启一个 Activity 时，生命周期执行的方法是：onCreate→onStart（可见，不可交互）→onResume（可见，可交互）。
- 单击 Back 键关闭一个 Activity 时，生命周期执行的方法是：onPause（部分可见，不可交互）→onStop（完全不可见）→onDestroy（销毁）。
- 当开启一个新的 Activity（以对话框形式），新的 Activity 把后面的 Activity 盖住一部分时，后面的 Activity 的生命周期执行的方法是：onPause（部分可见，不可交互）。注意，指定 Activity 以对话框的形式显示，需要在 Activity 节点追加以下主题 Android：theme="@android:style/Theme.Dialog"。
- 当把新开启的 Activity（以对话框形式）关闭时，后面的 Activity 的生命周期执行的方法是：onResume（可见，可交互）。
- 当开启一个新的 Activity，把后面的 Activity 完全盖住时，生命周期执行的方法是：onPause→onStop（完全不可见）。
- 当把新开启的 Activity（完全盖住）关闭时，生命周期执行的方法是：onRestart→onStart→onResume（可见，可交互）。

实际工作中常用的方法及应用场景有：①onResume，即可见，可交互，在该方法中可进行刷新数据操作；②onPause，即可见，但是不能响应用户操作，在该方法中可进行操作暂停；③onCreate，即初始化布局及一些大量的数据；④onDestroy，即释放数据，节省内存。

Activity 的生命周期如图 4-1 所示。

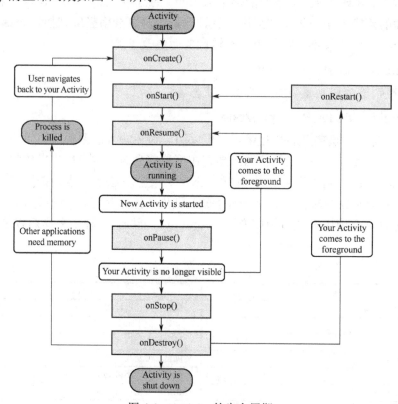

图 4-1　Activity 的生命周期

4.1.2　Intent

　　Intent 就是对将要执行的操作的抽象描述。Intent 支持 Android 设备上任意两个可用的应用程序组件之间的交互，无论它们是哪个应用程序的一部分。这就把一个相互独立的组件集合变成一个互联的系统。

　　Intent 最常用的一个用法是显式地（通过制定要加载的类）或隐式地（通过请求对一组数据执行某个动作）启动新的 Activity。

　　Intent 也可以用来在系统范围内广播消息。任何应用程序都可以注册一个广播接收器来监听和响应这些广播的 Intent。这样就可以基于内部的、系统的或第三方的应用程序事件，创建事件驱动的应用程序。

　　Android 使用广播 Intent 来公布系统事件，如网络连接状态或电池电量的改变。本地 Android 应用程序（如拨号程序和 SMS 管理器）可以简单地注册监听特定的广播 Intent 的组件，如来电或接收 SMS 消息并做出响应。使用 Intent 来传播动作——甚至在同一个应用程序内——是一种基于 Android 的设计原则。它鼓励组件之间的分离，允许无缝地替换应用程序元素。它还提供了一个简单的用于扩展功能模型的基础。[26]

　　Android 通常使用 Intent 实现一个界面到另一个界面的移动，它代表一个应用程序想要做什么。Intent 数据结构中最重要的两个部分是行动（Action）和数据（Data）。其中，Action 的典型值包括 MAIN、VIEW、PICK、EDIT 等，Data 则表示 URI。例如，要查看某个人的联系信息，可以使用 VIEW 操作创建一个 Intent，并将数据设置为表示该人的 URI。当 Intent 在 Android 组件间传递时，组件如果想告知 Android 系统自己能够响应和处理哪些 Intent，就需要用到 IntentFilter 对象。顾名思义，IntentFilter 对象负责过滤组件无法响应和处理的 Intent，只将组件关心的 Intent 接收进来进行处理。IntentFilter 实行"白名单"管理，即只列出组件期望接收的 Intent，但 IntentFilter 只会过滤隐式 Intent，显式的 Intent 则会直接传送到目标组件。Android 组件可以有一个或多个 IntentFilter，各 IntentFilter 之间相互独立，只需要其中一个验证通过即可。除了用于过滤广播的 IntentFilter 可以在代码中创建外，其他 IntentFilter 必须在 AndroidManifest.xml 文件中声明。

4.1.3　Service

　　Service 是一个长期运行的代码，运行时不需要用户界面。一个很好的例子就是媒体播放器从播放列表中播放歌曲。在媒体播放器的应用程序中，可能会有一个或多个允许用户选择歌曲并开始播放的 Activity。但是，音乐播放本身不应该由一个 Activity 来处理，因为即使在导航到新的屏幕之后,用户也期望音乐继续播放。在这种情况下,媒体播放器 Activity 可以使用 Context.startService()在后台运行来启动 Service，以保证歌曲的正常播放,此时系统将持续运行 Service 播放歌曲，直到完成。另外，还可以使用 Context.bindService()方法连接到 service（并在尚未运行的情况下启动它），通过 Service 的接口与 Service 进行通信。对于音乐 Service，可允许暂停、倒带等。

4.1.4　Broadcast

　　Broadcast 是 Android 中的四大组件之一，当执行电量低、开机、锁屏等操作都会发送一个广播。广播被分为两种不同的类型："普通广播（Normal broadcasts）"和"有序广播（Ordered broadcasts）"。普通广播是完全异步的，可以在同一时刻（逻辑上）被所有广播接收者接收到，消息传递的效率比较高；其缺点是：接收者不能将处理结果传递给下一个接收者，并且无法终止广播 Intent 的传播。有序广播中，接收者按照接收者声明的优先级别（声明在 intent-filter 元素的 android:priority 属性中，其值越大优先级别越高，其取值范围为 -1000～1000。也可以调用 IntentFilter 对象的 setPriority()进行设置）依次接收广播。例如，A 的级别高于 B，B 的级别高于 C，则广播先传给 A，再传给 B，最后传给 C。A 得到广播后，可以往广播里存入数据，当广播传给 B 时，B 可以从广播中得到 A 存入的数据。Context.sendBroadcast()发送的是普通广播，所有订阅者都有机会获得并进行处理。Context.sendOrderedBroadcast()发送的是有序广播，系统会根据接收者声明的优先级别按顺序逐个执行接收到的事件，前面的接收者有权终止广播（BroadcastReceiver.abortBroadcast()），如果广播被前面的接收者终止，后面的接收者就再也无法获取到广播。对于有序广播，前面的接收者可以将处理结果存放进广播 Intent，然后传给下一个接收者。[27]

4.1.5　Binder

　　Binder 通信是一种 Client-Server 的通信结构，从表面上来看，是 Client 通过获得一个 Server 的代理接口对 Server 进行直接调用。实际上，代理接口中定义的方法与 Server 中定义的方法是一一对应的。Client 调用某个代理接口中的方法时，代理接口的方法会将 Client 传递的参数打包成 Parcel 对象。代理接口将该 Parcel 发送给内核中的 Binder Driver。Server 会读取 Binder Driver 中的请求数据，如果是发送给自己的，则解包 Parcel 对象，处理并将结果返回。整个调用过程是一个同步过程，在 Server 处理时，Client 会被阻塞。[28]

4.1.6　Permission

　　随着 Android 6.0 的发布及普及，开发人员所要改变且适应的就是新版本 SDK 带来的一些新变化。首先需要关注的是权限机制的变化。从 Android 6.0 版本开始，在安装应用时，该应用无法取得任何权限。相反，在使用应用的过程中，若某个功能需要获取某个权限，系统会弹出一个对话框，显式地由用户决定是否将该权限赋予应用。只有得到了用户的许可，该功能才可以被使用。需要注意的是，赋予权限的对话框并不会自动弹出，需要由开发者手动调用。若程序调用的某个方法需要用户赋予相应的权限，而此时该权限并未被赋予，程序就会抛出异常并崩溃。除此之外，用户还可以在任何时候，通过设置中的应用管理撤销赋予过的权限。

当开发一个应用时，如果不配置 uses-permission，则这个应用几乎什么也做不了，系统没有给应用赋予默认的基础权限。任何操作权限都必须在 AndroidManifest.xml 文件中声明。这里的权限又分为以下 4 种。

普通权限（normal permission）：即使拥有了该类权限，用户的隐私数据被泄露或篡改的风险也很小。

敏感权限（dangerous permission）：与普通权限相反，一旦某个应用获取了该类权限，用户的隐私数据就面临被泄露或篡改的风险。例如，READ_CONTACTS 权限就属于敏感权限。

签名权限（signature permission）：该类权限只对拥有相同签名的应用开放，如手机 QQ 程序自定义了一个 permission，微信要去访问 QQ 的某个数据时，必须拥有该权限，这样手机 QQ 在自定义该权限时可以在权限标签中加入 android：protectionLevel= "signature"。然后在微信和手机 QQ 发布时采用相同的签名，这样微信就可以申请访问手机 QQ 中的某类开放数据了。即使其他程序知道了这个开放数据的接口，也在 manifest 注册申请了权限，但由于签名不同，还是无法访问的。

系统签名权限（signatureOrSystem permission）：与 signature permission 类似，但它不仅要求签名相同，还要求是同类的系统级应用，一般手机厂商开发的预制应用才会用到该类权限。

4.1.7 Manifest

AndroidManifest.xml 是每个 Android 程序中必需的文件，它位于整个项目的根目录中。我们每天都在使用这个文件，并往里面配置程序运行所必需的组件、权限，以及一些相关信息。AndroidManifest.xml 是 Android 应用的入口文件，它描述了 package 中暴露的组件（activities、services 等）、它们各自的实现类、各种能被处理的数据和启动位置。它除了能声明程序中的 Activities、Content Providers、Services 和 Intent Receivers 外，还能指定 permissions 和 instrumentation（安全控制和测试）[29]。接下来介绍 Manifest 的属性。

1. xmlns:android

定义 Android 命名空间，使得 Android 中的各种标准属性能在文件中使用，该空间提供了大部分元素中的数据。其格式一般为 http://schemas.android.com/apk/res/android。

2. package

指定本应用内 Java 主程序包的包名，它也是一个应用进程的默认名称。

3. sharedUserId

表明数据权限。因为默认情况下 Android 给每个 APK 分配一个唯一的 UserID，所以默认禁止不同 APK 访问共享数据。若要共享数据，一是可以采用 Share Preference 方法，二是可以采用 sharedUserId，将不同 APK 的 sharedUserId 都设为相同的，这些 APK 之间就可以互相共享数据了。

4. sharedUserLabel

一个共享的用户名。它只有在设置了 sharedUserId 属性的前提下才会有意义。

5. versionCode

是给设备程序识别版本（升级）用的。它必须是一个 interger 值，代表 App 更新过多少次，如第一版一般为 1，之后若要更新版本就设置为 2、3 等。

6. versionName

这个名称是给用户看的。可以将 App 版本号设置为 1.1，将后续更新版本设置为 1.2、2.0 等。

7. installLocation

安装参数，是 Android 2.2 中的一个新特性。installLocation 有 3 个值可以选择：preferExternal、auto、internalOnly。选择 preferExternal，系统会优先考虑将 APK 安装到 SD 卡上（当然最终用户可以选择为内部 ROM 存储，如果 SD 卡存储已满，也会安装到内部 ROM 上）。选择 auto，系统将会根据存储空间自己去适应。选择 internalOnly，则必须将其安装到内部才能运行。

一个 AndroidManifest.xml 中必须含有一个 Application 标签，这个标签声明了每一个应用程序的组件及其属性（如 icon、label、permission 等）。[29]

4.2 Android 工程结构

通过以下例子来认识开发应用的工程结构。

1. Android 工程

Android 工程目录如图 4-2 所示。

Android 工程中的文件详解如下。

--manifests：其中 AndroidManifest.xml 为 App 的配置信息。

--java：主要为源代码和测试代码。

--res：主要是资源目录，存储所有的项目资源。

  --drawable：存储 XML 文件。

  -*dpi 表示存储分辨率的图片，用于适配不同的屏幕。

   --mdpi：320×480。

   --hdpi：480×800、480×854。

   --xhdpi：至少 960×720。

图 4-2　Android 工程目录

--xxhdpi：1280×720。

　　--layout：存储布局文件。

　　--mipmap：存储原声图片资源。

　　--values：存储 App 引用的一些值。

　　--colors.xml：存储一些颜色的样式。

　　--dimens.xml：存储一些公用的 dp 值。

　　--strings.xml：存储引用的 string 值。

　　--styles.xml：存储 App 需要用到的一些样式。

--Gradle Scripts：build.gradle 为项目的 gradle 配置文件。

2．Project 工程

Project 工程目录如图 4-3 所示。

build：系统生成的文件目录，最后生成的 apk 文件就在这个目录中，这里是 app-debug.apk。

src：项目的源代码，其中 android test 为测试包，main 里为主要的项目目录和代码，test 为单元测试代码。

3．Packages 工程

Packages 工程目录如图 4-4 所示。

　　图 4-3　Project 工程目录　　　　　　图 4-4　Packages 工程目录

第 5 章 Android 应用用户界面设计

5.1 用户界面介绍

Android 应用程序的基本功能单元为 Activity 类的一个对象，主要功能为界面显示、单击事件处理等，其界面常使用 View（视图组件）和 ViewGroup（视图容器组件）来配置 XML 文件样式进行设计，而事件主要有按钮事件、触屏事件和对列表显示、文本框显示、图形显示等进行处理的高级控件的监听（就是对用户输入和网络接收的事件做出响应）。

5.1.1 Android 基本布局知识

Android 生成界面有三种形式：通过用户所在界面的接口生成；直接编写 Java 代码生成；用 XML 文件配置 Android 控件生成。

Android 应用程序的用户界面是由 View 和 ViewGroup 对象来组建的，二者均拥有独立的扩展子类，但 View 和 ViewGroup 又同属于 View 类的子类，开发者一般采用二者结合的形式进行手机界面设计。XML 文件只能用于界面布局设计和属性设置，不能用于事件处理，并且 Andriod 的界面组件一般没有太多变化，因此实现代码大都采用 Google 提供的代码命名模式，由单词字面意义就可基本了解其 UI 组件作用，开发者没有必要另起炉灶。

5.1.2 View 视图组件

View 类是 Android 中最基本的一个 UI 类，大多数高级 UI 组件都是继承 View 类而来的，如后面介绍的 TextView、EditView、Button、List、RadioButton、CheckBox 等。它的作用就是在屏幕上划分出一块矩形区域，并在其中进行特性化处理，设置符合需要的属性，如一般可以规划界面布局、改变区域的颜色、进行绘图、设置滚动条、设置测距和屏幕区域内用户需要表现的按键及手势，还可以在这个区域内设置相应的响应函数，进行事件处理等。后面将结合实例进行讲述。

在实际应用中，作为一个基类，View 类为 widget 服务，widget 是用于绘制交互屏幕元素的完全现实的子类，可以用它来构建用户界面 UI，实现多种多样的界面效果，满足用户和市场的需要。在编程过程中，Activity Java 文件中必须使用 import Android.widget.*** 进行包导入，如采用 import Android.widget.Button 导入按钮包文件，采用 import Android.widget.TextView 导入文本框包文件。

5.1.3　ViewGroup 视图容器组件

ViewGroup 是继承于 View 的子类，也属于一个抽象类。任意一个具体的 ViewGroup 对象都是 Android.view.ViewGroup 的实例体现。ViewGroup 是 View 的容器，负责添加 View 及对界面进行布局，体现了 Android 用户界面元素 UI 组件的多样性。因此，ViewGroup 将复杂的界面元素构建为一个独立的实体。实际应用中，作为一个基类，ViewGroup 为 Layout（布局）服务，Layout 是一组提供界面元素通用类型的完全现实的子类，可为 View 构建一个结构。

5.1.4　Layout 布局组件及其参数

Layout 是 ViewGroup 的实现服务，主要有五种形式，如表 5-1 所示。

表 5-1　Layout 布局

关 键 字	作 用	主 要 参 数
<LinearLayout [参数]> …组件… </LinearLayout>	线性布局，分为 Vertical（水平）、Horizontal（垂直）	layout_width, layout_height, background, padding
<RelativeLayout [参数]> …组件… </RelativeLayout>	相对布局	layout_width, layout_height, background
<FrameLayout [参数]> …组件… </FrameLayout>	框架布局	layout_width, layout_height, background, padding
<TableLayout [参数]> …组件… </TableLayout>	表格布局	layout_width, layout_height, background, padding, stretchColumns
<AbsoluteLayout [参数]> …组件… </AbsoluteLayout>	绝对布局	layout_width, layout_height, layout_x, layout_y, background, padding

各个布局模式中的参数可以不同。下面几行一般是要求的，若没有则为系统默认。

```
xmlns:android=http://schemas.android.com/apk/res/android  //此行是固定的
android:orientation="vertical"  //对于线性布局是必需的，其他布局中不需要此行代码
android:layout_width="fill_parent"  // fill_parent 可以改为 wrap_content
android:layout_height="fill_parent"  // fill_parent 可以改为 wrap_content
```

```
android:background="@drawable/bg1"//引用drawable文件夹中的图片(图片名称为bg1)
```
作为整个布局的背景

其他属性可以根据需要增加,如背景色、背景图片等,在此不再赘述。下面分别对以上 5 种布局进行详细介绍。

5.1.5　界面布局

1. AbsoluteLayout(绝对布局)

AbsoluteLayout(绝对布局)是指把多种组件放在同一个界面上,而组件彼此有精确的定位坐标。常见的方法是使用坐标实现定位,该布局方式适用于各种复杂的界面。运行结果如图 5-1 所示(完整源代码:AndroidDevelopment\Chapter5\Section5_1\AbsoluteLayout\ color1)。

2. LinearLayout(线性布局)

LinearLayout(线性布局)是最为常用的布局之一,只能为水平或垂直形式,通常采用垂直形式(完整源代码:Android Development\Chapter5\Section5_2\MixedLayout\Layout)。

布局文件代码 Layout/res/layout/main.xml 的主要内容如下:

```xml
<?xml version="1.0" encoding="utf-8"?>
<LinearLayout xmlns:android="http://schemas.android.com/apk/res/android"
    //若把 vertical 改为 horizontal 就是水平
    android:orientation="vertical"
    //设置背景图片
    android:background="@drawable/pgy2"
    //fill_parent 表示宽度为父容器的宽度,Wrap_content 表示内容长度为宽度
    android:layout_width="fill_parent"
    android:layout_height="fill_parent">
    <Button android:id="@+id/button0"
        android:layout_width="fill_parent"
        android:layout_height="wrap_content"
        android:text="FrameLayout 实现布局" />
    <Button android:id="@+id/button1"
        android:layout_width="fill_parent"
        android:layout_height="wrap_content"
        android:text="文字输入框:RelativeLayout 实现布局" />
    <Button android:id="@+id/button2"
        android:layout_width="fill_parent"
        android:layout_height="wrap_content"
        android:text="LinearLayout 和 RelativeLayout 共同实现布局" />
    <Button android:id="@+id/button3"
        android:layout_width="fill_parent"
        android:layout_height="wrap_content"
        android:text="用户登录:TableLayout 实现布局" />
</LinearLayout>
```

运行结果如图 5-2 所示。

说明：<?xml version="1.0" encoding="utf-8"?>中的 version=" 1.0"是指此文件用的 XML 版本是 1.0，encoding="UTF-8"是指此文件用的字符集是 UTF-8；android:background="@drawable/pggz"是设置整个线性布局背景为图片。

3．FrameLayout（框架布局）

FrameLayout（框架布局）直接在屏幕上开辟出一块空白的区域，当往里面添加控件时，会默认把它们放到这块区域的左上角，即所有控件都在左上角对齐。框架布局没有任何定位方式，该布局的大小由控件中最大的子控件决定，如果其中的控件相同，则同一时刻只能看到最上面的那个控件。在一个帧布局中显示一张图片的源代码如下所示（完整源代码：AndroidDevelopment\Chapter5\ Section5_2\MixedLayout\Layout）。

布局文件代码 Layout/res/layout/ activity_frame_layout.xml 的主要内容如下：

```xml
<?xml version="1.0" encoding="utf-8"?>
<FrameLayout android:id="@+id/left"
    xmlns:android="http://schemas.android.com/apk/res/android"
    android:layout_width="fill_parent"
    android:layout_height="fill_parent">
    <ImageView
        android:id="@+id/photo"
        android:layout_width="wrap_content"
        android:layout_height="wrap_content"
        android:src="@drawable/zw3"/>
</FrameLayout>
```

运行结果如图 5-3 所示。

图 5-1　AbsoluteLayout（绝对布局）运行结果

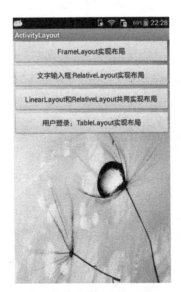

图 5-2　LinearLayout（线性布局）运行结果

图 5-3　FrameLayout（框架布局）运行结果

说明：只使用一个 ImageView 组件用于图片显示，android:src="@drawable/zw3"表示 drawable 文件夹中的图片（图片名称为 zw3），前面的标签 android: src 指定为图片源，不作为背景。

4. RelativeLayout（相对布局）

RelativeLayout（相对布局）首先指定一个目标，然后以它为基准进行相对布局（完整源代码：AndroidDevelopment\Chapter5\Section5_2\MixedLayout\Layout）。

布局文件代码 Layout\res\layout\relative_layout.xml 的主要内容如下：

```
<?xml version="1.0" encoding="utf-8"?>
<RelativeLayout
    xmlns:android="http://schemas.android.com/apk/res/android"
    android:layout_width="fill_parent"
    android:layout_height="wrap_content"
    android:background="@drawable/sy3"
    android:padding="10dip">
    <TextView android:id="@+id/label"
        android:layout_width="fill_parent"
        android:layout_height="wrap_content"
        android:textSize="18sp"
        android:textColor="#292421"
        android:text="请输入用户名：" />
    <!--这个EditText放置在上边id为label的TextView的下边-->
    <EditText android:id="@+id/entry"
        android:layout_width="fill_parent"
        android:layout_height="wrap_content"
        android:background="@android:drawable/editbox_background"
        android:layout_below="@id/label" />
    <!--取消按钮和容器的右边齐平，并且设置左边的边距为10dip-->
    <Button android:id="@+id/cancel"
        android:layout_width="wrap_content"
        android:layout_height="wrap_content"
        android:layout_below="@id/entry"
        android:layout_alignParentRight="true"
        android:layout_marginLeft="10dip"
        android:text="取消" />
    <!--确定按钮在取消按钮的左侧，并且和取消按钮的高度齐平-->
    <Button android:id="@+id/ok"
        android:layout_width="wrap_content"
        android:layout_height="wrap_content"
        android:layout_toLeftOf="@id/cancel"
        android:layout_alignTop="@id/cancel"
```

```
        android:text="确定" />
</RelativeLayout>
```

实验结果如图 5-4 所示。

5. LinearLayout 和 RelativeLayout 互助使用实例

下面讨论如何将 LinearLayout（线性布局）和 RelativeLayout（相对布局）结合使用以实现如图 5-5 所示的界面，该界面实现包括两个文件——界面布局 XML 文件和 Java 文件。该界面实际上由左右两个布局文件组成，相对布局需要使用相对变量来实现，即在 ActivityLayout.java 文件中用变量先后绑定左右两个相对布局文件后进行使用。

问题：布局文件 right.xml 和 left.xml 都使用了 RelativeLayout（相对布局），均没有提到 LinearLayout，为什么要说成 LinearLayout（线性布局）和 RelativeLayout（相对布局）互助使用？其实可以理解为把两个 RelativeLayout 相对布局文件在 ActivityLayout.java 文件中进行 LinearLayout 水平线性布局设置，具体就是在 ActivityLayout.java 文件中使用代码 layoutMain.setOrientation 完成水平线性布局设置，详细可参考 ActivityLayout.java 文件（完整源码：AndroidDevelopment\Chapter5\Section5_2\MixedLayout\Layout）。

布局文件代码 **Layout/res/layout/left.xml** 的主要内容如下：

```
<TextView android:id="@+id/view1"
    android:background="@drawable/zw12"
    android:layout_width="160px"
    android:layout_height="260px"
    android:text="左边第 1 项"
    android:textColor="#0B1746"
    android:textSize="25sp"
    android:gravity="center"
    android:layout_alignParentTop="true"
    android:layout_alignParentLeft="true" />
<TextView android:id="@+id/view2"
    android:background="@drawable/zw13"
    android:layout_width="160px"
    android:layout_height="260px"
    android:layout_below="@id/view1"
    android:text="左边第 2 项"
    android:textColor="#0B1746"
    android:textSize="25sp"
    android:gravity="center"/>
```

布局文件代码 **Layout/res/layout/right.xml** 的主要内容如下：

```
<TextView android:id="@+id/right_view1"
    android:background="@drawable/zw14"
    android:layout_width="160px"
    android:layout_height="260px"
```

```
            android:text="右边第1项"
            android:textColor="#0B1746"
            android:textSize="25sp"
            android:gravity="center"/>
    <TextView android:id="@+id/right_view2"
            android:background="@drawable/zw15"
            android:layout_width="160px"
            android:layout_height="260px"
            android:layout_below="@id/right_view1"
            android:text="右边第2项"
            android:textColor="#0B1746"
            android:textSize="25sp"
            android:gravity="center"/>
```

实验结果如图 5-5 所示。

图 5-4 RelativeLayout（相对 　　图 5-5 LinearLayout 和 RelativeLayout
　　　布局）运行结果　　　　　　　　　　互助使用运行结果

6．TableLayout（表格布局）

TableLayout（表格布局）是以表格的形式来表示数据的，下面以图 5-6 所示的一个简单登录界面为例学习如何使用表格布局。该登录界面包括 3 行，每行有两个单元格。若使用表格布局实现该界面，可以具体使用 TableRow 来实现其中的行，每行都可以自由添加组件，一般为按钮和文本框。例如，第一行的 TableRow 中包括 Textview 和 EditText 两个组件，第二行的 TableRow 中也包括 Textview 和 EditText 两个组件，第三行的 TableRow 中包括两个 Button 组件（完整源代码：AndroidDevelopment\Chapter5\Section5_2\MixedLayout\Layout）。

布局文件代码 Layout/res/layout/ activitytablelayout.xml 的主要内容如下：

```xml
<?xml version="1.0" encoding="utf-8"?>
<TableLayout xmlns:android="http://schemas.android.com/apk/res/android"
    android:layout_width="fill_parent"
    android:layout_height="fill_parent"
    android:background="@drawable/sy7"
    android:stretchColumns="1">
    <TableRow>
        <TextView android:text="用户名:"
            android:textStyle="bold"
            android:textColor="#BC8F8F"
            android:gravity="right"
            android:textSize="20sp"
            android:padding="42dip"
            android:alpha="0.8"/>
        <EditText android:id="@+id/username"
            android:padding="3dip"
            android:alpha="0.8"
            android:scrollHorizontally="true" />
    </TableRow>
    <TableRow>
        <TextView android:text="密码:"
            android:textStyle="bold"
            android:textColor="#BC8F8F"
            android:gravity="right"
            android:textSize="20sp"
            android:padding="40dip"
            android:alpha="0.8"/>
        <EditText android:id="@+id/password"
            android:password="true"
            android:padding="1dip"
            android:alpha="0.8"
            android:scrollHorizontally="true" />
    </TableRow>
    <TableRow android:gravity="left">
        <Button android:id="@+id/login"
            android:alpha="0.8"
            android:text="登录"/>
        <Button android:id="@+id/cancel"
            android:alpha="0.8"
            android:text="取消" />
    </TableRow>
```

```
</TableLayout>
```
运行结果如图 5-6 所示。

图 5-6　TableLayout（表格布局）运行结果

7．常数设置文件

文件代码 Layout\res\values\strings.xml 的主要内容如下：

```xml
<resources>
    <string name="hello">Hello World, ActivityLayoutTest</string>
    <string name="app_name">this is xml layout test</string>
    <drawable name="red">#7f00</drawable>
    <drawable name="blue">#770000ff</drawable>
    <drawable name="green">#7700ff00</drawable>
    <drawable name="yellow">#77ffff00</drawable>
    <drawable name="GRAY">#ff888888</drawable>
</resources>
```

8．系统自动生成的布局默认启动文件

文件代码 Layout\AndroidManifest.xml 的主要内容如下：

```xml
<?xml version="1.0" encoding="utf-8"?>
<manifest xmlns:android="http://schemas.android.com/apk/res/android"
    package="com.Layout" android:versionCode="1"
    android:versionName="1.0.0">
    <application android:icon="@drawable/icon"
        android:label="@string/app_name">
        <activity android:name=".ActivityMain"
            android:label="程序布局主界面：ActivityMain">
            <intent-filter>
                <action android:name="android.intent.action.MAIN" />
```

```xml
            <category
                android:name="android.intent.category.LAUNCHER"/>
        </intent-filter>
    </activity>
    <!--每添加一个 Activity 就必须按下面的格式注册一个 Activity 文件,否则出
    错。注意文件名和设置的主题显示名称的写法,android:label="混合 Layout 实
    现布局"给这个 Activity 贴一个标签,而在其他文件中设置主题名称-->
    <activity android:name=".ActivityLayout"
        android:label="混合 Layout 实现布局">
    </activity>
    <activity android:name=".ActivityRelativeLayout"
        android:label="RelativeLayout 实现布局">
    </activity>
    <activity android:name=".ActivityFrameLayout"
        android:label="FrameLayout 实现布局">
    </activity>
    <activity android:name=".ActivityTableLayout"
        android:label="TableLayout 实现布局">
    </activity>
</application>
</manifest>
```

9. 系统中 src 源文件夹的 Activity 文件

ActivityMain.java 的内容如下:

```java
package com.Layout;
import android.app.Activity;
import android.content.Intent;
import android.os.Bundle;
import android.view.View;
import android.view.View.OnClickListener;
import android.widget.Button;
public class ActivityMain extends Activity {
    OnClickListener listener0 = null;
    OnClickListener listener1 = null;
    OnClickListener listener2 = null;
    OnClickListener listener3 = null;
    Button button0;
    Button button1;
    Button button2;
    Button button3;
    /** Called when the activity is first created. */
    @Override
    public void onCreate(Bundle savedInstanceState) {
        super.onCreate(savedInstanceState);
```

```java
            listener0 = new OnClickListener() {
                public void onClick(View v) {
                    Intent intent0 = new Intent(ActivityMain.this,
                            ActivityFrameLayout.class);
                    setTitle("FrameLayout");
                    startActivity(intent0);
                }
            };
            listener1 = new OnClickListener() {
                public void onClick(View v) {
                    Intent intent1 = new Intent(ActivityMain.this,
                            ActivityRelativeLayout.class);
                    startActivity(intent1);
                }
            };
            listener2 = new OnClickListener() {
                public void onClick(View v) {
                    setTitle("ActivityLayout");
                    Intent intent2 = new Intent(ActivityMain.this,
                            ActivityLayout.class);
                    startActivity(intent2);

                }
            };
            listener3 = new OnClickListener() {
                public void onClick(View v) {
                    setTitle("TableLayout");
                    Intent intent3 = new Intent(ActivityMain.this,
                            ActivityTableLayout.class);
                    startActivity(intent3);
                }
            };
            setContentView(R.layout.main);
            button0 = (Button) findViewById(R.id.button0);
            button0.setOnClickListener(listener0);
            button1 = (Button) findViewById(R.id.button1);
            button1.setOnClickListener(listener1);
            button2 = (Button) findViewById(R.id.button2);
            button2.setOnClickListener(listener2);
            button3 = (Button) findViewById(R.id.button3);
            button3.setOnClickListener(listener3);
        }
    }
```

ActivityFrameLayout.java 的内容如下：

```java
package com.Layout;
import android.app.Activity;
import android.os.Bundle;
public class ActivityFrameLayout extends Activity {
    /** Called when the activity is first created. */
    @Override
    public void onCreate(Bundle savedInstanceState) {
        super.onCreate(savedInstanceState);
        setContentView(R.layout.activityframelayout);
    }
}
```

ActivityLayout.java 的内容如下：

```java
package com.Layout;
import android.app.Activity;
import android.content.Context;
import android.os.Bundle;
import android.view.LayoutInflater;
import android.widget.LinearLayout;
import android.widget.RelativeLayout;
public class ActivityLayout extends Activity {
    /** Called when the activity is first created. */
    @Override
    public void onCreate(Bundle savedInstanceState) {
        super.onCreate(savedInstanceState);
        LinearLayout layoutMain = new LinearLayout(this);
        layoutMain.setOrientation(LinearLayout.HORIZONTAL);
        setContentView(layoutMain);
        LayoutInflater inflate = (LayoutInflater) getSystemService(
        Context.LAYOUT_INFLATER_SERVICE);
        RelativeLayout layoutLeft = (RelativeLayout) inflate.inflate(
        R.layout.left, null);
        RelativeLayout layoutRight = (RelativeLayout) inflate.inflate(
        R.layout.right, null);
        RelativeLayout.LayoutParams relParam = new RelativeLayout.
        LayoutParams(
                RelativeLayout.LayoutParams.WRAP_CONTENT,
                RelativeLayout.LayoutParams.WRAP_CONTENT);
        layoutMain.addView(layoutLeft,160,520);
        layoutMain.addView(layoutRight, relParam);
    }
}
```

ActivityTableLayout.java 的内容如下：

```java
public class ActivityTableLayout extends Activity {
    /** Called when the activity is first created. */
```

```java
        @Override
        public void onCreate(Bundle savedInstanceState) {
            super.onCreate(savedInstanceState);
            setContentView(R.layout.activitytablelayout);
        }
    }
```

10. 系统自动生成的 R.java

R.java 的主要内容如下：

```java
    package com.Layout;
    public final class R {
        public static final class attr {
        }
        public static final class drawable {
            public static final int GRAY=0x7f02000c;
            public static final int bg=0x7f020000;
            public static final int bg1=0x7f020001;
            public static final int blue=0x7f02000d;
            public static final int green=0x7f02000e;
            public static final int icon=0x7f020002;
            public static final int jz1=0x7f020003;
            public static final int pgy2=0x7f020004;
            public static final int red=0x7f02000f;
            public static final int sy3=0x7f020005;
            public static final int sy7=0x7f020006;
            public static final int yellow=0x7f020010;
            public static final int zw12=0x7f020007;
            public static final int zw13=0x7f020008;
            public static final int zw14=0x7f020009;
            public static final int zw15=0x7f02000a;
            public static final int zw3=0x7f02000b;
        }
        public static final class id {
            public static final int button0=0x7f05000b;
            public static final int button1=0x7f05000c;
            public static final int button2=0x7f05000d;
            public static final int button3=0x7f05000e;
            public static final int cancel=0x7f050004;
            public static final int entry=0x7f050003;
            public static final int label=0x7f050002;
            public static final int left=0x7f050000;
            public static final int login=0x7f050008;
            public static final int ok=0x7f050005;
```

```java
            public static final int password=0x7f050007;
            public static final int photo=0x7f050001;
            public static final int right=0x7f05000f;
            public static final int right_view1=0x7f050010;
            public static final int right_view2=0x7f050011;
            public static final int username=0x7f050006;
            public static final int view1=0x7f050009;
            public static final int view2=0x7f05000a;
        }
        public static final class layout {
            public static final int activityframelayout=0x7f030000;
            public static final int activityrelativelayout=0x7f030001;
            public static final int activitytablelayout=0x7f030002;
            public static final int left=0x7f030003;
            public static final int main=0x7f030004;
            public static final int right=0x7f030005;
        }
        public static final class string {
            public static final int app_name=0x7f040000;
            public static final int hello=0x7f040001;
        }
    }
```

说明：R.java 是系统自动生成的文件，目录为 app→build→generated→source→r→debug →com.Layout→R.java。没有必要的话最好不要修改它，否则会编译错误。开发者编译文件时已经自动把所有的变量、常量、颜色和字符串常数等都记录在该文件内，并且定义为静态的、不可更改的对象。

5.1.6 事件处理的简单介绍

事件处理就是用户与 UI（图形界面）交互时所触发的动作，其本质就是函数调用。函数是在 Activity 的 Java 文件中使用的，可以自定义函数实现，其中既可以使用现成的系统函数，也可以直接使用单独的系统函数，完成触发操作。使用系统函数时，主要应了解系统函数传递的参数，但这里需要注意的是 Android 系统的部分函数功能是固定的，这些函数是无参数的，如 gettext()；读取文本。

5.2　Android 颜色的基本用法和介绍

Android 系统有如表 5-2 所示的 12 种常见的颜色。

表 5-2 常见的颜色

常数	颜色
Color.BLACK	黑色
Color.BLUE	蓝色
Color.CYAN	青绿色
Color.DKGRAY	灰黑色
Color.GRAY	灰色
Color.GREEN	绿色
Color.LTGRAY	浅灰色
Color.MAGENTA	紫红色
Color.RED	红色
Color.TRANSPARENT	透明色
Color.WHITE	白色
Color.YELLOW	黄色

这些颜色常数通常定义在 android.graphics.Color 中,如表 5-3 所示。

表 5-3 颜色类型定义

类型	常数	值	色码
int	BLACK	-16 777 216	0xff000000
int	BLUE	-16 776 961	0xff0000ff
int	CYAN	-16 711 681	0xff00ffff
int	DKGRAY	-12 303 292	0xff444444
int	GRAY	-7 829 368	0xff888888
int	GREEN	-16 711 936	0xff00ff00
int	LTGRAY	-3 355 444	0xffcccccc
int	MAGENTA	-65 281	0xffff00ff
int	RED	-65 536	0xffff0000
int	TRANSPARENT	0	0x00000000
int	WHITE	-1	0xffffffff
int	YELLOW	-256	0xffffff00

下面举例说明:使用 Color.MAGENTA 指定文本的颜色为紫红色,需要调用组件。

用法一:

```
mTextView02.setTextColor(Color.MAGENTA);
```

注意:需要在包含文件中加上 import android.graphics.Color 包才行。

用法二:

```
android:textColor="@drawable/darkgray"//设置文字颜色
android:background="@drawable/white"  //设置背景色
```

将定义好的颜色代码(color code)以 drawable 的名称(name)存放于 resources 中,这是 Android 开发程序的通常做法。如同字符串常数一样,颜色也可以在程序中先被定义。在 Value 文件夹下使用 drawable 的 resource 的定义方法:

```
<drawable name="color_name">color_value</drawable>
```
定义好的 drawable name 常数必须存放于 res/values 下面，进行封装，以作为资源取用，但定义好的背景颜色并非只能当作"默认"颜色声明使用，在程序的事件里可以通过程序来更改，如以下程序所示：

```
Resources resources = getBaseContext().getResources();
Drawable HippoDrawable = resources.getDrawable(R.drawable.white);
TextView tv = (TextView)findViewByID(R.id.text);
tv.setBackground(HippoDrawable);
```

举例说明：改变窗口背景色，通过使用 drawable 定义颜色实现（完整源代码：Android-Development\Chapter5\Section5_1\AbsoluteLayout\color1）。

color1\src\com\color1\color1.java 的代码如下：

```
package com.color1;
import android.app.Activity;
import android.os.Bundle;
import android.content.res.Resources;
import android.graphics.Color;
import android.graphics.drawable.Drawable;
import android.widget.TextView;
public class color1 extends Activity
{
    private TextView mTextView01;
    private TextView mTextView02;
/** Called when the activity is first created. */
    @Override
    public void onCreate(Bundle savedInstanceState)
    {
      super.onCreate(savedInstanceState);
      setContentView(R.layout.main);
      mTextView01 = (TextView) findViewById(R.id.myTextView01);
      mTextView01.setText("这是采用 Drawable 常数法设置文本框的背景色。");
      Resources resources = getBaseContext().getResources();
      Drawable HippoDrawable = resources.getDrawable(R.drawable.white);
      mTextView01.setBackgroundDrawable(HippoDrawable);
      mTextView02 = (TextView) findViewById(R.id.myTextView02);
      mTextView02.setTextColor(Color.YELLOW);
    }
}
```

程序说明：在开头一定要包含所有的系统类。下面三行代码为设置文本框的背景颜色：

```
Resources resources = getBaseContext().getResources();
Drawable HippoDrawable = resources.getDrawable(R.drawable.white);
mTextView01.setBackgroundDrawable(HippoDrawable);
```

但是这种设置方法非常复杂，不如在 XML 文件中定义组件时直接用常数法设置背景颜色更简单明了，这里只是演示一下这种方法，读者可以自行选择使用何种方法。

mTextView02.setTextColor(Color.YELLOW)；虽然只能设置文字颜色，但是灵活便捷，推荐使用。其中 R.Java 文件不用修改，它会在修改 XML 文件时自动生成。

布局文件代码 color1\res\layout\main.xml 的主要内容如下：

```xml
<?xml version="1.0" encoding="utf-8"?>
<AbsoluteLayout
    android:id="@+id/widget35"
    android:layout_width="fill_parent"
    android:layout_height="fill_parent"
    android:background="@drawable/GRAY"
    xmlns:android="http://schemas.android.com/apk/res/android"
>

<TextView
    android:id="@+id/widget28"
    android:layout_width="wrap_content"
    android:layout_height="wrap_content"
    android:text="@string/str_id"
    android:textColor="@drawable/white"
    android:layout_x="61px"
    android:layout_y="69px"
>
</TextView>
<TextView
    android:id="@+id/widget29"
    android:layout_width="wrap_content"
    android:layout_height="wrap_content"
    android:text="@string/str_pwd"
    android:textColor="@drawable/white"
    android:layout_x="61px"
    android:layout_y="158px"
>
</TextView>
<EditText
    android:id="@+id/widget31"
    android:layout_width="120dip"
    android:layout_height="wrap_content"
    android:textSize="18sp"
    android:layout_x="114px"
    android:layout_y="57px"
>
</EditText>
<EditText
    android:id="@+id/widget30"
```

```xml
        android:layout_width="120dip"
        android:layout_height="wrap_content"
        android:textSize="18sp"
        android:password="true"
        android:layout_x="112px"
        android:layout_y="142px"
    >
</EditText>
<TextView
    android:id="@+id/myTextView01"
    android:layout_width="260dip"
    android:layout_height="40dip"
    android:layout_x="30px"
    android:layout_y="242px"
    android:textColor="@drawable/darkgray"
    android:text="@string/str_textview01"
/>
<TextView
    android:id="@+id/myTextView02"
    android:layout_width="300dip"
    android:layout_height="wrap_content"
    android:layout_x="12px"
    android:layout_y="342px"
    android:textSize="20sp"
    android:text="@string/str_textview02"
/>
</AbsoluteLayout>
```

程序说明：该布局主体采用<AbsoluteLayout>……</AbsoluteLayout>实现绝对定位，常见的是使用坐标实现，比较适用于各种复杂的布局。

android:layout_x="61px"和 android:layout_y="69px"这两行是用坐标绝对定位 UI 组件的位置，需要注意的是坐标原点在左上角，坐标值的单位一般为 dip 或 sp 等，通常使用 dip，根据屏幕分辨率来规定像素，以便适用于各种机型。

android:background="@drawable/GRAY"就是调用颜色常数，更改屏幕背景颜色。

android:id="@+id/myTextView02"组件都定义了 id 号，便于重复使用。

android:text="@string/str_id"用于显示文字的常数引用，也是常见的方法。

android:textColor="@drawable/white"用于文字颜色的常数引用，很简洁，值得推荐。

color1\res\values\color.xml 主要用于颜色常数的定义：

```xml
<?xml version="1.0" encoding="utf-8"?>
<resources>
<drawable name="darkgray">#808080FF</drawable>
<drawable name="white">#FFFFFFFF</drawable>
<drawable name="GRAY">#ff888888</drawable>
```

```xml
    <drawable name="red">#ffff0000</drawable>
  </resources>
```
color1\res\values\strings.xml 主要用于一些常数文本的定义：
```xml
<?xml version="1.0" encoding="utf-8"?>
<resources>
    <string name="hello">Hello World, color1</string>
    <string name="app_name">color1</string>
    <string name="str_id">用户名</string>
    <string name="str_pwd">密码</string>
    <string name="str_textview01"></string>
    <string name="str_textview02">这是采用 graphics.Color 设置文本颜色，直接在 Java 文件中动态更改文字颜色
    mTextView02.setTextColor(Color.YELLOW);</string>
</resources>
```

5.3 基本组件介绍和应用

Android 开发中对于组件的 id 号的定义是随意的，前提是不重复，因为无论在哪个文件里定义的组件，系统最终都会把各个组件的 id 号集成在 R.java 文件里，需要使用或指定动作时一般提供其对应的 id 号即可。当然也可以不定义，但是这样不方便以后使用或指定某个特定的组件，因此推荐都定义 id 号。

常用的 UI 组件有 widgit、menu、Listview、Dialog、Toast 和 Notification。

常用的 widgit 基本组件包括 TextView（文本框）、EditText（文本编辑框）、Button（按钮）、RadioButton（单选框）、ScrollView（滚动视图）等，后面一一介绍。

为了便于学习，本节采用布局项目和颜色设置项目中的源代码对组件进行介绍，以加强读者对各个组件之间的搭配使用。另外，将演示一个包括全部基本 widget 组件的代码，其中大多数组件为了方便理解，使用单独的代码来加以说明。读者可以用 Android Studio 打开项目边学习边修改，这样效率更高一些。

5.3.1 Widget 组件

下面学习 Android 系统的 Widget 组件及它们的实现方法（完整源代码：Android-Development\Chapter5\Section5_3\WidgetComponent\Widget）。

1. TextView（文本框）

TextView（不可编辑的文本框）的具体属性如下：
```xml
<?xml version="1.0" encoding="utf-8"?>
<LinearLayout xmlns:android="http://schemas.android.com/apk/res/android"
```

```
        android:orientation="vertical"
        android:layout_width="fill_parent"
        android:layout_height="fill_parent"
        >
    <TextView
        android:id="@+id/text_view"
        android:layout_width="fill_parent"
        android:layout_height="wrap_content"
        android:layout_x="20dip"
        android:layout_y="40dip"
        android:textSize="22sp"
        android:textColor="#8A2BE2"
        android:padding="10dip"
        android:background="@drawable/sy11"
        android:text="这里是文本框 TextView，展示你的 freestyle！"
        />
</LinearLayout>
```

运行结果如图 5-7 所示。

图 5-7　文本框运行结果

特别地，TextView 的属性可以在 Java 文件中设置。TextView 的两种属性设置方法如表 5-4 所示。

表 5-4　TextView 的两种属性设置方法

Java 文件中的 myTextView 对象调用	XML 文件中设置 TextView 属性
setText(string)	android:text="string"
setTextSize(float sp)	android:textSize="20sp"
setTextColor(int color)	android:textColor="#ff00ffff"
setBackgroundResource(int resid)	android:background="Colorvalue"
setWidth(int pixels)	android:height="200dip"
setHeight(intpixels)	android:width="200dip"

例如：
```
myTextView.setTextColor(Color.CYAN);
myTextView.setTextSize(screenHeight);
myTextView.setText(string);
android:background="#ff00ffff"
android:background="@drawable/bg1"
```

2. EditText（文本编辑框）

布局项目中相对布局文件里的文本的基本设置如下：
```
<EditText android:id="@+id/edit_text"
    android:layout_width="fill_parent"
    android:layout_height="70sp"
    android:hint="请在这里输入内容"
/>
<Button android:id="@+id/get_edit_view_button"
    android:layout_width="wrap_content"
    android:layout_height="wrap_content"
    android:layout_gravity="right"
    android:text="提交内容"
    android:alpha="0.9"
/>
```
运行结果如图 5-8 所示。

3. Button（按钮）

Button 是 Android 系统中最常用的组件，通常需要对其增加监听事件。一般在 XML 布局文件中设计 Button 的属性，包括文字信息、长宽、颜色等；然后在 Java 文件中定义 Button 并指向布局中的 Button，并为其设置监听（完整源代码：AndroidDevelopment\Chapter5\Section5_2\MixedLayout\Layout）。

以图 5-9 所示的界面为例，需要在界面上显示按钮，并为该按钮增加监听事件。首先在布局文件 main.xml 中建立一个 id 号为 button0 的按钮，该按钮上的文字信息为"FrameLayout 实现布局"，其代码如下：
```
<Button android:id="@+id/button0"
    android:layout_width="fill_parent"
    android:layout_height="wrap_content"
    android:text="FrameLayout 实现布局"
/>
```
然后在 Java 文件中定义按钮并增加监听时间。具体的流程如下：首先定义 Button 对象 button0，并使用 findViewById 方法根据 id 号指向布局文件中的按钮；然后使用 setOnClickListener 方法对该按钮增加单击监听事件 listener0。具体实现代码如下：
```
setContentView(R.layout.main);  // 显示布局文件
```

```
button0 = (Button) findViewById(R.id.button0);// 定义按钮对象
button0.setOnClickListener(listener0);// 对按钮增加监听事件
```
其中使用 setContentView(R.layout.main)方法绑定 main.xml 布局文件和 Java 文件，从而显示 main.xml 布局文件。

进一步，可以具体定义监听事件 listener0，如定义当单击按钮时，界面跳转至另外的一个框架布局界面，则该监听事件的代码如下：
```
listener0 = new OnClickListener() {
    public void onClick(View v) {
        Intent intent0 = new Intent(ActivityMain.this,
        ActivityFrameLayout.class);
        setTitle("FrameLayout");
        startActivity(intent0);
    }
};
```
运行结果如图 5-9 所示。

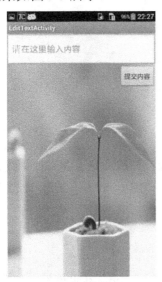
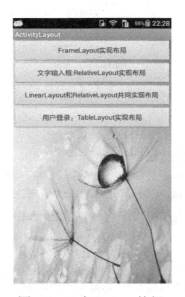

图 5-8 文本编辑框运行结果　　　　图 5-9 4 个 Button 按钮

4．RadioButton（单选框）

单选框是指在多个选项中每次只能选中其中一个选项。例如，对于"性别"，用户只能选择选项"男"或选项"女"。单选框的实现过程是：通过 RadioGroup 组件将多个 RadioButton 组件组合起来，其中一个 RadioButton 组件实现一个选项。图 5-10 所示的单选框的实现基本代码如下：
```
<?xml version="1.0" encoding="utf-8"?>
<LinearLayout xmlns:android="http://schemas.android.com/apk/res/android"
    android:layout_width="fill_parent"
    android:layout_height="fill_parent"
    android:orientation="vertical">
```

```xml
<RadioGroup
    android:id="@+id/menu"
    android:layout_width="fill_parent"
    android:layout_height="wrap_content"
    android:background="@drawable/zw11"
    android:checkedButton="@+id/lunch"
    android:orientation="vertical">
    <RadioButton
        android:id="@+id/breakfast"
        android:text="breakfast"
        android:textSize="20sp"/>
    <RadioButton
        android:id="@id/lunch"
        android:text="lunch"
        android:textSize="20sp"/>
    <RadioButton
        android:id="@+id/dinner"
        android:text="dinner"
        android:textSize="20sp"/>
    <RadioButton
        android:id="@+id/all"
        android:text="all"
        android:textSize="20sp"/>
</RadioGroup>
</LinearLayout>
```

外层是一个 RadioGroup 组件，包括该组件的高、宽、id 号等属性；内层包含多个 RadioButton，每个 RadioButton 对应一个选项。RadioGroup 用于把所有的 RadioButton 封装起来，使得每次只有一个 RadioButton 组件被选中。

运行结果如图 5-10 所示。

5. Spinner（下拉列表）

下拉列表组件的实现方式相对麻烦一些，有两种实现方式。其总体界面如图 5-11 所示。
（1）第一种：首先在 XML 文件里定义界面如下：

```xml
<Spinner  android:id="@+id/spinner_1"
    android:layout_width="fill_parent"
    android:layout_height="wrap_content"
    android:drawSelectorOnTop="false"
/>
```

然后在 SpinnerActivity.java 中设置如下：

```java
private static final String[] mCountries = { "China" ,"Russia", "Germany",
                                    "Ukraine", "Belarus", "USA" };
    private void find_and_modify_view() {
        spinner_c = (Spinner) findViewById(R.id.spinner_1);
        allcountries = new ArrayList<String>();
```

```
        for (int i = 0; i < mCountries.length; i++) {
            allcountries.add(mCountries[i]);
        }
    }
```

用方法 find_and_modify_view()载入数据，定义一个 ArrayList<String>，遍历数组，把内容存在集合里，放在 ArrayAdapter（数组容器）对象里并指定位置，然后进行选择。用 SetDropDownViewResource 的方法调用 ArrayAdapter 对象。最后用 spinner 对象的 setAdapter 方法设置添加的内容。运行结果如图 5-12 所示。

图 5-10　单选框　　　　图 5-11　下拉列表　　　　图 5-12　XML 定义界面的下拉列表

（2）第二种：界面代码。首先尝试在文件夹 Values 下的 array.xml 文件里预先定义所有的选项数据：

```
<Spinner android:id="@+id/spinner_2"
    android:layout_width="fill_parent"
    android:layout_height="wrap_content"
    android:drawSelectorOnTop="false"
/>
<item>China2</item>
<item>Russia2</item>
<item>Germany2</item>
<item>Ukraine2</item>
<item>Belarus2</item>
<item>USA2</item>
</string-array>
</resources>
<?xml version="1.0" encoding="utf-8"?>
<resources>
<!-- Used in Spinner/spinner_2.java -->
<string-array name="countries">
spinner_2 = (Spinner) findViewById(R.id.spinner_2);
```

```
ArrayAdapter<CharSequence> adapter = ArrayAdapter.createFromResource(
this,R.array.countries,android.R.layout.simple_spinner_item);
adapter.setDropDownViewResource(android.R.layout.simple_spinner_dropdown_
item);
spinner_2.setAdapter(adapter);
```
然后在对应的 SpinnerActivity.java 里进行调用。

在 res/values/目录下的 array.xml 文件中通过 R.array.countries 找到数组元素，然后将 setDropDownViewResource()方法设置为 spinner_2 的内容，运行结果如图 5-13 所示。

6．AutoCompleteTextView（自动完成文本）

该组件需要把 COUNTRIES 数组的内容绑定进 ArrayAdapter<String>里，再把它与对应的 XML 布局文件中的组件联合起来：

```
ArrayAdapter<String> adapter = new ArrayAdapter (this,
android.R.layout.simple_dropdown_item_1line,new String[]{"someshing 某
物" ," someone 某人", "somebody 某人","sometimes 有时", "somewhat 有点",
"somehow 以某种方法" ,"somewhere 某地" ,"someones 某人", "somebodys 某人
","someplace 某地","same 相同的", "small 小的", "select 选择"});
AutoCompleteTextView textView = (AutoCompleteTextView) findViewById(R.
                                id.auto_complete);
textView.setAdapter(adapter);
```

对应的 XML 文件 AutoComplete.xml 的内容如下：

```
<AutoCompleteTextView
     android:id="@+id/auto_complete"
     android:layout_width="fill_parent"
     android:layout_height="wrap_content"
     />
```

当输入 so、some 时与其匹配的内容就会自动显示出来。运行结果如图 5-14 所示。

图 5-13　界面布局的下拉列表　　　　　图 5-14　自动完成文本

7. DatePicker（日期选择器）

先用 findViewById()绑定布局组件，再用 init 方法对日期进行初始化：

```
date_Picker.init( 2017, 9 , 10 , new DatePicker.OnDate ChangedListener() {
@Override
  public void onDateChanged(DatePicker view, int year, int monthOfYear,
  intdayOfMonth) {
      Calendar calendar = Calendar.getInstance();
      calendar.set(year, monthOfYear, dayOfMonth);
      SimpleDateFormat format = new SimpleDateFor mat("yyyy 年 MM 月 dd
      日 HH:mm");
      Toast.makeText(DatePickerActivity.this,format.format(calendar.
                        getTime()),Toast.LENGTH_SHORT).show();
   }
});
```

对应的 layout/date_picker.xml 的主要内容如下：

```
<DatePicker
    android:id="@+id/date_picker"
    android:layout_width="wrap_content"
    android:layout_height="wrap_content"
    android:calendarViewShown="false"
    android:layout_gravity="center"
    android:alpha="0.85"/>
```

运行结果如图 5-15 所示。

8. TimePicker（时间选择器）

先用 findViewById()绑定布局组件，再用 init 方法对时间进行初始化。
TimePickerActivity.java 文件的相关代码如下：

```
public void onCreate(Bundle savedInstanceState) {
      super.onCreate(savedInstanceState);
      setTitle("TimePickerActivity");
      setContentView(R.layout.time_picker);
      imePicker tp = (TimePicker)this.findViewById(R.id.time_picker);
      p.setIs24HourView(false);
}
```

对应的 layout/time_picker.xml 的主要内容如下：

```
<TimePicker
    android:id="@+id/time_picker"
    android:layout_width="wrap_content"
    android:layout_height="wrap_content"
    android:layout_gravity="center"
/>
```

运行结果如图 5-16 所示。

9. SeekBar(拖动条)

在 Java 文件中直接启动 setContentView(R.layout.seek_bar)。

对应的 layout/seek_bar.xml 的主要内容如下:

```
<TextView
    android:layout_width="wrap_content"
    android:layout_height="wrap_content"
    android:layout_gravity="end"
    android:text="先定一个小目标!"
    android:textColor="#191970"
    android:textSize="23sp"/>
<SeekBar
    android:id="@+id/seek"
    android:layout_width="fill_parent"
    android:layout_height="wrap_content"
    android:max="200"
    android:thumb="@drawable/seeker"
    android:progress="100"/>
```

运行结果如图 5-17 所示。

图 5-15 日期选择器　　图 5-16 时间选择器　　图 5-17 拖动条

10. ScrollView(滚动视图)

当主界面要显示的内容超过一个屏幕时,可以在 XML 文件布局容器外面套上 <ScrollView>…</ScrollView>,代码的主要内容如下:

```
<?xml version="1.0" encoding="utf-8"?>
<ScrollView xmlns:android="http://schemas.android.com/apk/res/android"
    android:layout_width="fill_parent"
    android:layout_height="wrap_content"
    android:background="@drawable/sy6">
<LinearLayout xmlns:android="http://schemas.android.com/apk/res/android"
```

```xml
android:orientation="vertical"
android:layout_width="fill_parent"
android:layout_height="fill_parent">
<Button android:id="@+id/text_view_button"
    android:layout_width="260sp"
    android:layout_height="50sp"
    android:textSize="15sp"
    android:alpha="0.75"
    android:layout_gravity="center"
    android:text="文本框[ TextView ]" />
<Button android:id="@+id/edit_view_button"
    android:layout_width="260sp"
    android:layout_height="50sp"
    android:textSize="15sp"
    android:alpha="0.75"
    android:layout_gravity="center"
    android:text="编辑框[ EditView ]" />
<Button android:id="@+id/radio_group_button"
    android:layout_width="260sp"
    android:layout_height="50sp"
    android:textSize="15sp"
    android:alpha="0.75"
    android:layout_gravity="center"
    android:text="单选框[RadioGroup]" />
<Button android:id="@+id/check_box_button"
    android:layout_width="260sp"
    android:layout_height="50sp"
    android:textSize="15sp"
    android:alpha="0.75"
    android:layout_gravity="center"
    android:text="复选框[ CheckBox ]" />
<Button android:id="@+id/auto_complete_button"
    android:layout_width="260sp"
    android:layout_height="50sp"
    android:textSize="15sp"
    android:alpha="0.75"
    android:layout_gravity="center"
    android:text="自动提示[AutoCompleteTextView]" />
<Button android:id="@+id/spinner_button"
    android:layout_width="260sp"
    android:layout_height="50sp"
    android:textSize="15sp"
```

```xml
        android:alpha="0.75"
        android:layout_gravity="center"
        android:text="下拉列表[Spinner]" />
    <Button android:id="@+id/time_picker_button"
        android:layout_width="260sp"
        android:layout_height="50sp"
        android:textSize="15sp"
        android:alpha="0.75"
        android:layout_gravity="center"
        android:text="时间选择器[TimePicker]" />
    <Button android:id="@+id/date_picker_button"
        android:layout_width="260sp"
        android:layout_height="50sp"
        android:textSize="15sp"
        android:alpha="0.75"
        android:layout_gravity="center"
        android:text="日期选择器[DatePicker]" />
    <Button android:id="@+id/progress_bar_button"
        android:layout_width="260sp"
        android:layout_height="50sp"
        android:textSize="15sp"
        android:alpha="0.75"
        android:layout_gravity="center"
        android:text="进度条[ProgressBar]" />
    <Button android:id="@+id/seek_bar_button"
        android:layout_width="260sp"
        android:layout_height="50sp"
        android:textSize="15sp"
        android:alpha="0.75"
        android:layout_gravity="center"
        android:text="拖动条[SeekBar]" />
    <Button android:id="@+id/rating_bar_button"
        android:layout_width="260sp"
        android:layout_height="50sp"
        android:textSize="15sp"
        android:alpha="0.75"
        android:layout_gravity="center"
        android:text="评分组件[RatingBar]" />
    <Button android:id="@+id/image_view_button"
        android:layout_width="260sp"
        android:layout_height="50sp"
        android:textSize="15sp"
```

```xml
            android:alpha="0.75"
            android:layout_gravity="center"
            android:text="图片视图[ImageView]" />
        <Button android:id="@+id/image_show_button"
            android:layout_width="260sp"
            android:layout_height="50sp"
            android:textSize="15sp"
            android:alpha="0.75"
            android:layout_gravity="center"
            android:text="切换图片[ImageSwitcher Gallery]" />
        <Button android:id="@+id/grid_view_button"
            android:layout_width="260sp"
            android:layout_height="50sp"
            android:textSize="15sp"
            android:alpha="0.75"
            android:layout_gravity="center"
            android:text="网格视图[GridView]" />
        <Button android:id="@+id/tab_demo_button"
            android:layout_width="260sp"
            android:layout_height="50sp"
            android:textSize="15sp"
            android:alpha="0.75"
            android:layout_gravity="center"
            android:text="切换标签[TabView]" />
    </LinearLayout>
</ScrollView>
```

运行结果如图 5-18 所示。

11. ProgressBar（进度条）

在 Java 中需要用 setContentView(R.layout.progress_bar) 启动布局文件，其对应的 layout/progress_bar.xml 的主要内容如下：

```xml
<ProgressBar
    android:id="@+id/progress_bar"
    android:layout_width="wrap_content"
    android:layout_height="wrap_content"/>
<ProgressBar android:id="@+id/progress_horizontal"
    style="?android:attr/progressBarStyleHorizontal"
    android:layout_width="200dip"
    android:layout_height="wrap_content"
    android:max="100"
    android:progress="50"
    android:secondaryProgress="75" />
```

运行结果如图 5-19 所示。

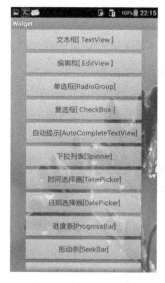
图 5-18 滚动视图

图 5-19 进度条

12. RatingBar（评分星级组件）

在某些应用中需要用到评分或评比，用五角星表示等级。Java 中设有启动布局文件，其对应的 layout/rating_bar.xml 的主要内容如下：

```
<RatingBar android:id="@+id/rating_bar"
    android:layout_width="wrap_content"
    android:layout_height="wrap_content"/>
<TextView
    android:layout_width="wrap_content"
    android:layout_height="wrap_content"
    android:singleLine="false"
    android:layout_gravity="center"
    android:text="超给力！"
    android:textColor="#8A2BE2"
    android:textSize="20sp"/>
```

运行结果如图 5-20 所示。

13. ImageView（图片视图）

此组件用于展示一张指定的图片。其对应的 layout/image_view.xml 的主要内容如下：

```
<ImageView
    android:id="@+id/imageview"
    android:src="@drawable/qiangwei"
    android:layout_width="280px"
    android:layout_height="410px"/>
<TextView
    android:layout_width="wrap_content"
    android:layout_height="wrap_content"
```

```
android:text="完整图片展开"
android:textColor="#0B1746"
android:textSize="30sp"/>
```

运行结果如图 5-21 所示。

图 5-20　评分星级组件　　　　图 5-21　图片视图

14．GridView（网格视图）

在 GridViewActivity.java 中定义 getView()方法，创建一个 imageView，并设定图片 imageView.setImageResource（mThumbIds[position]）：

```
public class ImageAdapter extends BaseAdapter {
private Context mContext;
    public ImageAdapter(Context c) {
        mContext = c;
    }
    public int getCount() {
        return mThumbIds.length;
    }
    public Object getItem(int position) {
        return null;
    }
    public long getItemId(int position) {
        return 0;
    }
    public View getView(int position, View convertView, ViewGroup parent) {
        ImageView imageView;
        if (convertView == null) {
            imageView = new ImageView(mContext);
```

```java
            imageView.setLayoutParams(new GridView.LayoutParams(85, 85));
            imageView.setScaleType(ImageView.ScaleType.CENTER_CROP);
                imageView.setPadding(8, 8, 8, 8);
        } else {
            imageView = (ImageView) convertView;
        }
        imageView.setImageResource(mThumbIds[position]);
        return imageView;
    }
    // 关联到drawable文件夹里的图片源
    private Integer[] mThumbIds = {
            R.drawable.gridview1, R.drawable.gridview2,
            R.drawable.gridview3, R.drawable.gridview4,
            R.drawable.gridview5, R.drawable.gridview6,
            R.drawable.gridview7, R.drawable.gridview8,
            R.drawable.gridview9, R.drawable.gridview10,
            R.drawable.gridview11, R.drawable.gridview12,
            R.drawable.gridview13, R.drawable.gridview14,
            R.drawable.gridview, R.drawable.sample1,
            R.drawable.sample2, R.drawable.sample3,
            R.drawable.sample4, R.drawable.sample5,
            R.drawable.sample6, R.drawable.sample0
    };
}
```

对应的 layout/grid_view.xml 的主要内容如下：

```xml
<GridView xmlns:android="http://schemas.android.com/apk/res/android"
    android:id="@+id/grid_view"
    android:layout_width="fill_parent"
    android:layout_height="fill_parent"
    android:numColumns="auto_fit"
    android:verticalSpacing="10dp"
    android:horizontalSpacing="10dp"
    android:columnWidth="90dp"
    android:stretchMode="columnWidth"
    android:gravity="center"/>
```

运行结果如图 5-22 所示。

15. ImageSwitcher&Gallery［（画廊）切换图片］

com/Widget/ImageShowActivity.java 文件的主要内容如下：

```java
public void onCreate(Bundle savedInstanceState) {
    super.onCreate(savedInstanceState);
    requestWindowFeature(Window.FEATURE_NO_TITLE);
```

```java
    setContentView(R.layout.image_show);
    setTitle("ImageShowActivity");
    mSwitcher = (ImageSwitcher) findViewById(R.id.switcher);
    mSwitcher.setFactory(this);
    mSwitcher.setInAnimation(AnimationUtils.loadAnimation(this,
    android.R.anim.fade_in));
    mSwitcher.setOutAnimation(AnimationUtils.loadAnimation(this,
    android.R.anim.fade_out));

    Gallery g = (Gallery) findViewById(R.id.gallery);
    g.setAdapter(new ImageAdapter(this));
    g.setOnItemSelectedListener(this);
}
public void onItemSelected(AdapterView parent, View v, int position,
long id) {
    mSwitcher.setImageResource(mImageIds[position]);
}
public void onNothingSelected(AdapterView parent) {
}
public View makeView() {
    ImageView i = new ImageView(this);
    i.setBackgroundColor(0xFF000000);
    i.setScaleType(ImageView.ScaleType.FIT_CENTER);
    i.setLayoutParams(new ImageSwitcher.LayoutParams(LayoutParams.
    FILL_PARENT,LayoutParams.FILL_PARENT));
    return i;
}
private ImageSwitcher mSwitcher;
public class ImageAdapter extends BaseAdapter {
    public ImageAdapter(Context c) {
        mContext = c;
    }
    public int getCount() {
        return mThumbIds.length;
    }
    public Object getItem(int position) {
        return position;
    }
    public long getItemId(int position) {
        return position;
    }
    public View getView(int position, View convertView, ViewGroup
```

```java
            parent) {
        ImageView i = new ImageView(mContext);
        i.setImageResource(mThumbIds[position]);
        i.setAdjustViewBounds(true);
        i.setLayoutParams(new Gallery.LayoutParams(
        LayoutParams.WRAP_CONTENT, LayoutParams.WRAP_CONTENT));
        i.setBackgroundResource(R.drawable.picture_frame);
        return i;
    }
    private Context mContext;
}
private Integer[] mThumbIds = {
        R.drawable.sample_thumb_0, R.drawable.sample_thumb_1,
        R.drawable.sample_thumb_2, R.drawable.sample_thumb_3,
        R.drawable.sample_thumb_4, R.drawable.sample_thumb_5,
        R.drawable.sample_thumb_6, R.drawable.sample_thumb_7};
private Integer[] mImageIds = {
        R.drawable.sample_0, R.drawable.sample_1, R.drawable.sample_2,
        R.drawable.sample_3, R.drawable.sample_4, R.drawable.sample_5,
        R.drawable.sample_6, R.drawable.sample_7};
```

makeView 方法说明：ViewSwitcher.ViewFactory 接口里有一个 makeView 方法，makeView 方法中初始化了一个 ImageView 并设置其相关属性，最后将该值返回。ViewSwitcher 通过该方法初始化各种资源。

getView 方法说明：要显示整个图片，需要先对 BaceAdapter 进行封装，通过 getView() 方法来返回要显示的 ImageView。

对应的 layout/image_show.xml 的主要内容如下：

```xml
<ImageSwitcher
        android:id="@+id/switcher"
        android:layout_width="fill_parent"
        android:layout_height="fill_parent"
        android:layout_alignParentTop="true"
        android:layout_alignParentLeft="true" />
<Gallery android:id="@+id/gallery"
        android:background="#55000000"
        android:layout_width="fill_parent"
        android:layout_height="60dp"
        android:layout_alignParentBottom= "true"
        android:layout_alignParentLeft= "true"
        android:gravity="center_vertical"
        android:spacing="16dp" />
```

运行结果如图 5-23 所示。

图 5-22　网格视图　　　　图 5-23　（画廊）切换图片

16．TabHost（切换面板）

此组件先用 getTabHost() 方法获取一个 TabHost，然后进行模板绑定：

```
public void onCreate(Bundle savedInstanceState) {
    super.onCreate(savedInstanceState);
    setTitle("TabDemoActivity");
    TabHost tabHost = getTabHost();
    LayoutInflater.from(this).inflate(R.layout.tab_demo,tabHost.get
TabContentView(), true);
    tabHost.addTab(tabHost.newTabSpec("tab1").setIndicator("tab1")
.setContent(R.id.view1));
    tabHost.addTab(tabHost.newTabSpec("tab3").setIndicator("tab2")
.setContent(R.id.view2));
    tabHost.addTab(tabHost.newTabSpec("tab3").setIndicator("tab3")
.setContent(R.id.view3));
}
```

对应的 layout/tab_demo.xml 的主要内容如下：

```
<?xml version="1.0" encoding="utf-8"?>
<FrameLayout xmlns:android="http://schemas.android.com/apk/res/android"
    android:layout_width="fill_parent"
    android:layout_height="fill_parent"
    android:gravity="center">
<TextView android:id="@+id/view1"
    android:background="@drawable/sy6"
    android:layout_width="fill_parent"
    android:layout_height="fill_parent"
    android:singleLine="false"
    android:text="\n\n\n厉害了，我的 Android! "
```

```
                android:textColor="#D1EEEE"
                android:textSize="30sp"
                android:layout_gravity="center"
                />
        <TextView android:id="@+id/view2"
                android:background="@drawable/zw18"
                android:layout_width="fill_parent"
                android:layout_height="fill_parent"
                android:singleLine="false"
                android:text="\n\n\n        这样的开发工具\n          是极好的！"
                android:textColor="#FFFAFA"
                android:textSize="30sp"
                android:layout_gravity="center"/>
        <TextView android:id="@+id/view3"
                android:background="@drawable/zw19"
                android:layout_width="fill_parent"
                android:layout_height="fill_parent"
                android:singleLine="false"
                android:text="\n\n    我为Android Studio打Call！"
                android:textColor="#FFF5EE"
                android:textSize="30sp"
                android:layout_gravity="center"/>
    </FrameLayout>
```

运行结果如图 5-24 所示。

图 5-24　切换面板

5.3.2　ListView 列表

ListView 是 Android 系统中最常用的组件之一，它一般要与数据库配合使用，这里简单学习其界面实现。ListView 组件中的每个子项 Item 既可以是一个字符串，也可以是一个组合控件。简单来说,适配器就是 Item 数组,动态数组有多少个元素就生成多少个 Item;然后把适配器添加到 ListView 中,并显示出来(完整源代码:AndroidDevelopment\Chapter5\Section5_4\List\ListView)。

下面演示三种列表,并对其进行简单操作。

第一种 com/ListView/ListView1.java 的主要内容如下:

```java
public class ListView1 extends Activity {
    ListView listView;
    private String[] data = { "甘肃省","陕西省","吉林省","宁夏回族自治区",
        "辽宁省", "四川省", "西藏自治区", "青海省","安徽省", "湖北省","贵州省",
        "福建省" };
    @Override
    public void onCreate(Bundle savedInstanceState) {
        super.onCreate(savedInstanceState);
        listView = new ListView(this);
        listView.setAdapter(new ArrayAdapter<String>(this,
            android.R.layout.simple_list_item_1, data));
        listView.setAdapter(new ArrayAdapter<String>(this,
            android.R.layout.simple_list_item_single_choice, data));
        listView.setItemsCanFocus(true);
        listView.setChoiceMode(ListView.CHOICE_MODE_MULTIPLE);
        setContentView(listView);
    }
}
```

对应的布局组件如下:

```xml
<ListView android:id="@id/android:list"
    android:layout_width="fill_parent"
    android:layout_height="fill_ par ent"/>
```

在 Java 文件中定义数组,用 **ArrayAdapter** 实现显示列表功能。运行结果如图 5-25 所示。

图 5-25　Java 定义的 ListView 列表

代码的主要内容如下:

```java
public class ListView2 extends Activity {
    private List<Map<String, Object>> data;
    private ListView listView = null;
    public void onCreate(Bundle savedInstanceState) {
        super.onCreate(savedInstanceState);
        PrepareData();
        listView = new ListView(this);
        SimpleAdapter adapter = new SimpleAdapter(this, data,
        android.R.layout.simple_list_item_2, new String[]
        { "name","num" },
        new int[] { android.R.id.text1 , android.R.id.text2});
        listView.setAdapter(adapter);
        setContentView(listView);
        OnItemClickListener listener = new OnItemClickListener() {
        public void onItemClick(AdapterView<?> parent, View view,
        int position,long id) {
            setTitle(parent.getItemAtPosition (position).toString());
        }
        };
listView.setOnItemClickListener(listener);
    }
}
```

第二种 com/ListView/ListView2.java 的主要内容如下:

```java
private void PrepareData() {
    data = new ArrayList<Map<String, Object>>();
    Map<String, Object> item;
    item = new HashMap<String, Object>();
    item.put("name", "张三");
    item.put("num", "学号06333");
    data.add(item);
    item = new HashMap<String, Object>();
    item.put("name", "李四");
    item.put("num", "学号00023");
    data.add(item);
    item = new HashMap<String, Object>();
    item.put("name", "王五");
    item.put("num", "学号00012");
    data.add(item);
    item = new HashMap<String, Object>();
    item.put("name", "朱六");
    item.put("num", "学号00076");
```

```
            data.add(item);
            item = new HashMap<String, Object>();
            item.put("name", "阮七");
            item.put("num", "学号00054");
            data.add(item);
        }
```

运行结果如图 5-26 所示。

第三种 com/ListView/ListView3.java 的主要内容如下：

```
    private String[] data ={};
    public void onCreate(Bundle savedInstanceState) {
        super.onCreate(savedInstanceState);
        setContentView(R.layout.list3);
        setListAdapter(new ArrayAdapter<String>(this,
        android.R.layout.simple_list_item_1, data));
    }
     protected void onListItemClick(ListView listView, View v, int position,
    long id) {
        super.onListItemClick(listView, v, position, id);
        setTitle(listView.getItemAtPosition(position).toString());
    }
```

运行结果如图 5-27 所示。

图 5-26 Layout 定义显示

图 5-27 无数据时的提示

5.3.3 Notification 状态栏提示

Notification 是具体的状态栏通知对象，可以设置 icon、文字、提示声音、振动等参数。可选的设置：A ticker-text message（状态栏顶部提示消息），An alert sound（提示音），A vibrate setting（振动），A flashing LED setting（灯光）（完整源代码：AndroidDevelopment\Chapter5\Section5_5\Notification\notificationActivity）。

代码的主要内容如下：

```java
publicclass notificationActivity extends Activity{
    Buttonm_Button1, m_Button2, m_Button3, m_Button4;
    NotificationManager m_NotificationManager;
    Intent m_Intent;
    PendingIntent m_PendingIntent;
    Notification m_Notification;
    /** Called when the activity is first created. */
    @Override
    public void onCreate(Bundle savedInstanceState){
        super.onCreate(savedInstanceState);
        setContentView(R.layout.main);
        m_NotificationManager = (NotificationManager) getSystemService
        (NOTIFICATION_SERVICE);
        m_Button1 = (Button) findViewById(R.id.Button01);
        m_Button2 = (Button) findViewById(R.id.Button02);
        m_Button3 = (Button) findViewById(R.id.Button03);
        m_Button4 = (Button) findViewById(R.id.Button04);
        m_Intent = new Intent(notificationActivity.this,
        notificationActivity1. class);
        m_PendingIntent = PendingIntent.getActivity(notificationActivity.
        this, 0, m_Intent, 0);
        m_Notification = new Notification();
        m_Button1.setOnClickListener(new Button.OnClickListener() {
            public void onClick(View v){
                m_Notification.icon = R.drawable.img1;
                m_Notification.tickerText = "今天晴空万里，适宜户外活动";
                m_Notification.defaults = Notification.DEFAULT_SOUND;
                m_Notification.setLatestEventInfo(notificationActivity.
                this,"Button1", "Button提示", m_PendingIntent);
                m_NotificationManager.notify(0, m_Notification);
            }
        });
        m_Button2.setOnClickListener(new Button.OnClickListener() {
            public void onClick(View v){
                m_Notification.icon = R.drawable.img2;
                m_Notification.tickerText = "今天阴转小雨，外出带雨具";
                m_Notification.defaults = Notification.DEFAULT_VIBRATE;
                m_Notification.setLatestEventInfo(notificationActivity.
                this,"Button2", "Button2提示", m_PendingIntent);
                m_NotificationManager.notify(0, m_Notification);
            }
        });
        m_Button3.setOnClickListener(new Button.OnClickListener() {
```

```
            public void onClick(View v){
                m_Notification.icon = R.drawable.img3;
                m_Notification.tickerText = "今天大到暴雨,不宜外出";
                m_Notification.defaults = Notification.DEFAULT_LIGHTS;
                m_Notification.setLatestEventInfo(notificationActivity.
                    this, "Button3", "Button3 提示", m_PendingIntent);
                m_NotificationManager.notify(0, m_Notification);
            }
        });
        m_Button4.setOnClickListener(new Button.OnClickListener() {
            public void onClick(View v){
                m_Notification.icon = R.drawable.img4;
                m_Notification.tickerText = "今天雨夹雪,注意保暖";
                m_Notification.defaults = Notification.DEFAULT_ALL;
                m_Notification.setLatestEventInfo(notificationActivity.
                    this, "Button4", "Button4 通知",m_PendingIntent);
                m_NotificationManager.notify(0, m_Notification);
            }
        });
    }
}
```

运行结果如图 5-28 和图 5-29 所示,请注意上方的提示框显示栏。

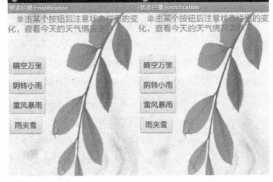

图 5-28　单击 button1 和 button2 的响应　　　图 5-29　单击 button3 和 button4 的响应

这个文件中的 4 种提示类型是相似的,不同之处在于手机提示时的动作不同,有响铃、振动、屏幕发亮、发亮振动 4 种方式。

为了触发一个 Notification,可使用 NotificationManager 的 notify 方法,将一个整数的 ID 和 Notification 对象传入,如下所示:

```
int notificationRef = 1;
notificationManager.notify(notificationRef, notification);
```

为了更新一个已经触发过的 Notification,可传入相同的 id(既可以传入相同的 Notification 对象,也可以传入一个全新的对象)。只要 id 相同,新的 Notification 对象会替换状态条图标和扩展的状态窗口的细节。

5.3.4 Toast 临时提示框

Toast 是一种信息提示方式。图 5-30 所示的 Toast 临时提示消息的具体实现如下：

```java
package com.ToastActivity;
import android.app.Activity;
import android.graphics.Color;
import android.graphics.Typeface;
import android.os.Bundle;
import android.view.View;
import android.widget.Button;
import android.widget.TextView;
import android.widget.Toast;
public class ToastActivity extends Activity {
    private TextView textview;
    @Override
    public void onCreate(Bundle savedInstanceState) {
        super.onCreate(savedInstanceState);
        setContentView(R.layout.main);
        String str = "#当收到短信时，我们会提示你信息内容！请点击 Toast 测试，注意下边的变化效果！";
        textview = (TextView)this.findViewById(R.id.textview1);
        textview.setTextSize(30);
        textview.setTypeface(Typeface.createFromAsset(getAssets(),
        "font/simkai.ttf"));
        textview.setTextColor(Color.YELLOW);
        textview.setText(str);
        Button button = (Button) findViewById(R.id.button1);
        button.setOnClickListener(new Button.OnClickListener() {
            public void onClick(View v){
                DisplayToast("你好,短信内容在这里显示,请及时查看！");
            }
        });
    }
    public void DisplayToast(String str) {
        Toast.makeText(this, str, Toast.LENGTH_SHORT).show();
    }
}
```

完整源代码：AndroidDevelopment\Chapter5\Section5_6\Toast\ToastActivity。

对应的 layout/main.xml 的主要内容如下：

```xml
<TextView
    android:id="@+id/textview1"
    android:layout_width="fill_parent"
    android:layout_height="wrap_content"
    android:text="@string/hello"
    android:textSize="25sp"
/>
<Button
    android:id="@+id/button1"
    android:layout_width="wrap_content"
    android:layout_height="wrap_content"
    android:text="测试 Toast 按钮"
    android:layout_gravity="center"
    android:textSize="18sp"
    android:alpha="0.8"
/>
```

运行结果如图 5-30 所示。

图 5-30　Toast 临时提示框

5.3.5　Dialog 对话框

定制对话框需要用到系统的类，这也是 Android 系统的惯例。对话框一般需要定义一个继承自 Dialog 的类，Dialog 是区别于 View 的另一个独立的类，它是直接从 Java.lang.Object 开始创建的，由 Avtivity 来维护（包括生成、保存、恢复）。

对于图 5-31 所示的示例，界面中有 4 个按钮，当单击每个按钮时均会弹出对应的对话框。下面详细分析这些对话框的实现过程（完整源代码：AndroidDevelopment\Chapter5\

Section5_7\Dialog\dialog)。

首先需要把 XML 文件中的 4 个 Button 组件与 Java 文件中的 4 个 Button 对象绑定起来：

```
Button button1 = (Button) findViewById(R.id.button1);
button1.setOnClickListener(new OnClickListener() {
    public void onClick(View v) {
        showDialog(DIALOG1);
    }
});
```

然后选择要调用的对话框的方法，针对不同 Dialog 的 id，buildDialog1 生成对应 id 的 Dialog：

```
protected Dialog onCreateDialog(int id) {
    switch (id) {
        case DIALOG1:return buildDialog1(dialog.this);
        case DIALOG2:return buildDialog2(dialog.this);
        case DIALOG3:return buildDialog3(dialog.this);
        case DIALOG4:return buildDialog4(dialog.this);
    }
    return null;
}
```

定义函数 buildDialog1()，该函数用于创建第一个对话框：

```
private Dialog buildDialog1(Context context) {
    AlertDialog.Builder builder = new AlertDialog.Builder(context);
    builder.setIcon(R.drawable.tishi);
    builder.setTitle(R.string.title1);
    builder.setPositiveButton(R.string.enter,
        newDialogInterface.OnClick Listener() {
        public void onClick(DialogInterface dialog, int whichButton) {
            setTitle(R.string.titleenter);
        }
    });
    builder.setNegativeButton(R.string.cancel,
        new DialogInterface.OnClickListener() {
        public void onClick(DialogInterface dialog, int whichButton) {
            setTitle(R.string.titlecancel);
        }
    });
    return builder.create();
}
```

其中，AlertDialog.Builder builder = new AlertDialog.Builder（context）表示动态生成一个 AlertDialog.Builder 对象，并开始构造 AlertDialog。注意 AlertDialog 是 Dialog 的一个子类，是 Android 系统中常用的对话框，可以对其设置主题和内容：添加组件，以丰富内容。由此可见，AlertDialog 类似于一个简单的 Activity。

setPositiveButton()和 setNegativeButton()两个函数分别用于对确定和取消按钮进行属性

设置（第一个参数）和事件监听（第二个参数）。最后使用 builder.create()创建一个设置完成的 Dialog。单击第一个按钮时弹出对话框的运行结果如图 5-31 所示。

对于第二个对话框，它的实现方式与第一个对话框基本一致，只是对话框里增加内容：
```
builder.setMessage(R.string.buttons2_msg);
```
下面增加一个用 setNeutralButton()方法定义的详细信息按钮进行监听：
```
builder.setNeutralButton(R.string.massage,
    new DialogInterface.OnClickListener() {
        public void onClick(DialogInterface dialog, int whichButton) {
            setTitle(R.string.massage);
        }
    });
```
单击第二个按钮时弹出对话框的运行结果如图 5-32 所示。

图 5-31　主页面及单击 button1 的响应　　　图 5-32　单击 button2 的响应

对于第三个对话框，它的实现方式与第一个对话框基本一致，只是在建立对话框之前先嵌入一个线性布局文件：
```
builder.setView(textEntryView);
```
View 定义：
```
LayoutInflater inflater = LayoutInflater.from(this);
final View textEntryView = inflater.inflate(R.layout.userentry, null);
```
单击第三个按钮时弹出对话框的运行结果如图 5-33 所示。

对于第四个对话框，其定义更直接明了：
```
private Dialog buildDialog4(Context context) {
    ProgressDialog dialog = new ProgressDialog(context);
    dialog.setTitle(R.string.loading);
    dialog.setMessage("请稍候……");
    return dialog;
}
```

单击第四个按钮时弹出对话框的运行结果如图 5-34 所示。

图 5-33　单击 button3 的响应　　　图 5-34　单击 button4 的响应

5.4　获取屏幕属性

在实际应用中，由于各种终端设备（手机）的屏幕尺寸有所不同，根据设备获得相应的屏幕属性可以得到更好的显示效果。对此，Android 系统提供了 DisplayMetrics 类来定义屏幕的一些属性，并且通过 getMetrics 方法来获取当前的 DisplayMetrics 属性（完整源代码：AndroidDevelopment\Chapter5\Section5_8\DisplayMetrics\getDisplay Metrics）。

代码的主要内容如下：

```
public class getDisplayMetrics extends Activity {
    TextView myTextView=null;
    public void onCreate(Bundle savedInstanceState) {
        super.onCreate(savedInstanceState);
        setContentView(R.layout.main);
        DisplayMetrics dm = new DisplayMetrics();
        getWindowManager().getDefaultDisplay().getMetrics(dm);
        int screenWidth = dm.widthPixels;
        int screenHeight = dm.heightPixels;
        myTextView = (TextView) findViewById(R.id.TextView01);
        myTextView.setText("\n 屏幕宽度："+screenWidth+"\n 屏幕高度："+
        screenHeight+"\n 屏幕大小："+screenWidth+"X"
         +screenHeight);
        myTextView.setTextColor(Color.WHITE);
        myTextView.setTypeface(Typeface.createFromAsset(getAssets(),
        "font/msyhbd.ttf"));
```

```
        myTextView.setTextSize(25);
    }
}
```
运行结果如图 5-35 所示。

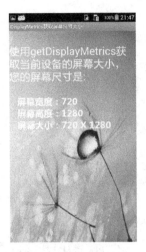

图 5-35　屏幕属性

第6章 Android 图形编程

6.1 Android 图形开发框架

用户界面设计是 Android 系统中不可忽视的一部分,利用 Android UI 开发自绘控件和进行游戏制作时,绘图基础是必不可少的。绘图开发必须在某个特定的框架下进行,下面介绍两种开发框架。

6.1.1 View 类开发框架

View 类是 Android 系统的一个超类,几乎包括了所有的屏幕类型,每个 View 中都有一个用于绘画的画布。用户还可以在应用开发中自定义 View 视图,让画布充分满足开发需求,如游戏、地图等。Android 系统中的 View 需要重新定义 onDraw 方法来实现界面的绘制及显示,用户自定义的视图可以分为普通的文本形式或较为复杂的 2D/3D 视图。

下面用屏幕上的一个视图颜色变化来说明 Android 的界面更新过程(完整源代码:AndroidDevelopment\Chapter6\Section6_1\ActivityView\ActivityViewkuangjia)。

代码的主要内容如下:

```
public class ActivityViewkuangjia extends Activity{
    private static final int REFRESH = 0x000001;
    private Viewkj mViewkj = null;
    /** Called when the activity is first created. */
    @Override
    public void onCreate(Bundle savedInstanceState){
        super.onCreate(savedInstanceState);
        getWindow().setBackgroundDrawableResource(R.drawable.zw7);
        this.mViewkj = new Viewkj(this);
        setContentView(mViewkj);
        new Thread(new GameThread()).start();
    }
    Handler myHandler  = new Handler() {
```

```
            public void handleMessage(Message msg){
                switch (msg.what){
                   case ActivityViewkuangjia.REFRESH:
                        mViewkj.invalidate();
                        break;
                }
                super.handleMessage(msg);
            }
        };
        class GameThread implements Runnable{
            public void run(){
                while (!Thread.currentThread().isInterrupted()){
                    try{
                        Thread.sleep(100);
                    }
                    catch (InterruptedException e){
                        Thread.currentThread().interrupt();
                    }
                    mViewkj.postInvalidate();
                }
            }
        }
```

运行结果如图 6-1 所示。

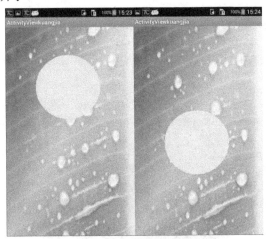

图 6-1 图形界面及按下键后的响应

6.1.2 SurfaceView 类开发框架

View 类开发框架在处理基本的绘图时较为常用，但是当需要开发更复杂多变的绘图软件时，Android 系统提供了一个专门的 SurfaceView 类来完成开发处理事项。例如，在高端游戏中利用栓缓冲显示，将复杂的背景、动画、人物等放在同一个画布（Canvas）上后，SurfaceView 可直接访问画布进行处理。Surface 是 Android 系统的一个重要概念，View 及其基本组件都要画在 Surface 上。由 Surface 创建的一个 Canvas 对象来管理 View 在 Surface 上的绘图操作。需要注意的是，使用 SurfaceView 绘图时要对其进行创建，状态改变时要进行监视，用完之后要进行销毁，这就要实现 SurfaceHolder.Callback 接口。[30]例如，对话裁剪、大小控制需要用 SurfaceHolder.来处理（完整源代码：AndroidDevelopment\Chapter6\Section6_2 \ActivitySurfaceView\ActivitySurfaceViewkuangjia）。

代码的主要内容如下：

```
public class ActivitySurfaceViewkuangjia extends Activity{
    SurfaceViewkj mSurfaceViewkj;
    public void onCreate(Bundle savedInstanceState){
        super.onCreate(savedInstanceState);
        mSurfaceViewkj = new SurfaceViewkj(this);
        setContentView(mSurfaceViewkj);
    }
    public boolean onTouchEvent(MotionEvent event){
        return true;
    }
    public boolean onKeyDown(int keyCode, KeyEvent event){
        return true;
    }
    public boolean onKeyUp(int keyCode, KeyEvent event){
        switch (keyCode){
        case KeyEvent.KEYCODE_DPAD_UP:
            mSurfaceViewkj.y-=3;
            break;
        case KeyEvent.KEYCODE_DPAD_DOWN:
            mSurfaceViewkj.y+=3;
            break;
        }
        return false;
    }
    public boolean onKeyMultiple(int keyCode, int repeatCount, KeyEvent event){
        return true;
    }
}
```

运行结果如图 6-2 所示。

图 6-2　矩形图形界面

6.2 Graphics 类

6.2.1 android.graphics.Color 类

Android 平台上表示颜色的方法有很多种，Color 提供了常规主要颜色的定义，如 Color.BLUE 和 Color.YELLOW 等，平时创建时主要使用以下静态方法：

```
static int argb(int alpha, int red, int green, int blue)   //构造一个包含透明对象的颜色
static int rgb(int red, int green, int blue)   //构造一个标准的颜色对象
static int parseColor(String colorString)   //解析一种颜色字符串的值，如传入 Color.BLACK
```

运行效果如图 6-3 所示。

图 6-3 实心图形

6.2.2 android.graphics.Paint 类

Paint 类可以理解为画笔、画刷，通常用到的方法如下：

```
void reset()    //重置
void setAntiAlias(boolean aa)   //是否抗锯齿，需要配合下面的语句来帮助消除锯齿使其边缘更平滑
void setFlags (Paint.ANTI_ALIAS_FLAG)
void setARGB(int a, int r, int g, int b) 或 void setColor(int color)   //均为设置 Paint 对象的颜色
 Shader  setShader(Shader shader)   //设置阴影，Shader 类是一个矩阵对象，如果为 NULL 将清除阴影，可以灵活使用
void setStyle(Paint.Style style)   //设置样式，一般为 FILL（填充）或 STROKE（凹陷效果）
void setTextSize(float textSize)   //设置字体大小
void setTextAlign(Paint.Align align)   //文本对齐方式
Typeface setTypeface(Typeface typeface)   //设置字体，通过 Typeface 方法可以加载 Android 内部的字体，一般为宋体。对于中文，部分 ROM 可以自己添加，如雅黑等
void setUnderlineText(boolean underlineText)   //是否设置下画线，一般需要配合 void setFlags (Paint.UNDERLINE_TEXT_FLAG) 方法[31]
```

运行效果如图 6-4 所示。

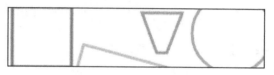

图 6-4 Paint 画笔效果

6.2.3 绘制几何图形

Android 系统中提供了一些基本的几何图形绘制。下面对常用的几种图形进行简单总结，如表 6-1 所示。

表 6-1 绘制几何图形的方法

基本的绘图函数	绘图说明
drawRect（rect1, mPaint） drawRect（float left, float top, float right, float bottom, Paint paint）	绘制矩形
drawCircle（float cx, float cy, float radius, Paint paint）	绘制圆形
drawOval（RectF oval, Paint paint）	绘制椭圆
drawPath（Path path, Paint paint）	绘制任意多边形
drawLine（float startX, float startY, float stopX, float stopY, Paint paint）	绘制直线
drawPoint（float x, float y, Paint paint）	绘制点

6.2.4 android.graphics.Canvas 类

Java 为我们的开发提供了 Canvas 类，而在 Android 平台上，Canvas 类的首要任务是管理绘制过程。将 Canvas 当作一块画布，我们可以在画布上绘制出想要的图形图像。这就牵涉到许多关于画布的属性，如颜色、尺寸、坐标等。下面先看一个 Paint、Color、Canvas 和几何绘图联合应用的实际例子，再了解 Canvas 的其他扩展应用。创建项目 Paint_Color，只需要使用两个 Java 文件即可（完整源代码：AndroidDevelopment \Chapter6\Section6_3\ Paint_Color_Canvas\Paint_Color）。

com/Paint_Color/PaintView.java 的主要内容如下：

```
package com.Paint_Color;
import android.content.Context;
import android.graphics.Canvas;
import android.graphics.Color;
import android.graphics.Paint;
import android.graphics.Path;
import android.graphics.Rect;
import android.graphics.RectF;
import android.view.KeyEvent;
import android.view.MotionEvent;
import android.view.View;
```

```java
public class PaintView extends View implements Runnable{
    private Paint mPaint = null;//声明画笔 Paint 对象
    public PaintView(Context context){
        super(context);
        mPaint = new Paint();   //动态构建对象
        new Thread(this).start();//开启线程
    }
    public void onDraw(Canvas canvas){
        super.onDraw(canvas);
        mPaint.setAntiAlias(true);// 设置画笔,取消锯齿效果
        canvas.clipRect(0, 0, 800, 1300);  //设置画布裁剪区域
        canvas.drawColor(Color.WHITE);// 设置画布的背景颜色
        mPaint.setStyle(Paint.Style.STROKE);{
            canvas.save();// 锁定画布
            canvas.rotate(15.0f);  //画布旋转 15 度角
            mPaint.setColor(Color.CYAN);
            canvas.drawRect(newRect(150,75,280,100), mPaint);
            canvas.restore();  //解除画布的锁定
            //另一种方法绘制矩形
            Rect rect1 = new Rect();   //定义矩形对象
            // 设置矩形大小及其定位
            rect1.left = 50;
            rect1.top = 100;
            rect1.bottom = 300;
            rect1.right = 390;
            mPaint.setColor(Color.MAGENTA);//修改画笔颜色
            canvas.drawRect(rect1, mPaint);
            //画一个圆
            mPaint.setColor(Color.GREEN);
            canvas.drawCircle(500, 160, 100, mPaint); // (圆心 x,圆心 y,半径 r,p)
            //画椭圆
            RectF rectf1 = new RectF();//定义椭圆对象
            //设置椭圆大小及其定位
            rectf1.left =300;
            rectf1.top = 400;
            rectf1.right = 150;
            rectf1.bottom = 600;
            mPaint.setColor(Color.LTGRAY);
            canvas.drawOval(rectf1, mPaint);    // 绘制椭圆
            // 绘制多边形
            Path path1 = new Path();
            //设置多边形的点
            path1.moveTo(600+5, 350-50);
```

```java
            path1.lineTo(600+45, 350-50);
            path1.lineTo(600+30, 420-50);
            path1.lineTo(600+20, 420-50);
            path1.close();//使这些点构成封闭的多边形
            mPaint.setColor(Color.GRAY);
            canvas.drawPath(path1, mPaint);// 绘制定义好的多边形
            mPaint.setColor(Color.BLUE);
            mPaint.setStrokeWidth(4);
            /* 绘制直线 */
            canvas.drawLine(80, 420, 600, 420, mPaint);
        }
        //############下面绘制实心几何体#######################
        mPaint.setStyle(Paint.Style.FILL);{
            //绘制实心矩形
            mPaint.setColor(Color.DKGRAY);
            canvas.drawRect(470, 500, 450, 450, mPaint);
            //绘制实心圆形(圆心x,圆心y,半径r,p)
            mPaint.setColor(Color.YELLOW);
            canvas.drawCircle(50, 200, 35, mPaint);
            //定义椭圆对象
            RectF rectf2 = new RectF();
            //设置椭圆大小
            rectf2.left = 200;
            rectf2.top = 220;
            rectf2.right = 260;
            rectf2.bottom = 170;
            mPaint.setColor(Color.LTGRAY);
            canvas.drawOval(rectf2, mPaint);   // 绘制椭圆
            // 绘制多边形
            Path path1 = new Path();
            //设置多边形的点
            path1.moveTo(150+5, 160+80-50);
            path1.lineTo(150+45, 160+80-50);
            path1.lineTo(150+30, 160+120-50);
            path1.lineTo(150+20, 160+120-50);
            path1.close(); //使这些点构成封闭的多边形
            mPaint.setColor(Color.GRAY);
            canvas.drawPath(path1, mPaint);//绘制这个多边形
            //绘制直线
            mPaint.setColor(Color.RED);
            mPaint.setStrokeWidth(3);
            canvas.drawLine(25, 260, 310, 260, mPaint);
        }
        // 触笔事件
```

```java
    public boolean onTouchEvent(MotionEvent event)
    {    return true;  }
    // 按键按下事件
    public boolean onKeyDown(int keyCode, KeyEvent event)
    {    return true;  }
    // 按键弹起事件
    public boolean onKeyUp(int keyCode, KeyEvent event)
    {    return false;    }
    public boolean onKeyMultiple(int keyCode, int repeatCount, KeyEvent
    event){      return true;      }
    //运行机制
    public void run(){
        while (!Thread.currentThread().isInterrupted()){
           try{
               Thread.sleep(100);
           }catch (Exception e){
           Thread.currentThread().interrupt();
           }
        postInvalidate();// 使用 postInvalidate 直接在线程中更新界面
      }
    }
```

com/Paint_Color/ShapeDraw.java 的主要内容如下：

```java
package com.Paint_Color;
import android.content.Context;
import android.graphics.Canvas;
import android.graphics.Color;
import android.graphics.Path;
import android.graphics.Rect;
import android.graphics.drawable.ShapeDrawable;
import android.graphics.drawable.shapes.OvalShape;
import android.graphics.drawable.shapes.PathShape;
import android.graphics.drawable.shapes.RectShape;
import android.view.View;
public class ShapeDraw extends View{
    ShapeDrawable mShapeDrawable    = null;//声明 ShapeDrawable 对象
    public ShapeDraw(Context context){
        super(context);
    }
    public void DrawShape(Canvas canvas){
        //实例化 ShapeDrawable 对象并说明是绘制一个矩形
        mShapeDrawable = new ShapeDrawable(new RectShape());
        //得到画笔 paint 对象并设置其颜色
        mShapeDrawable.getPaint().setColor(Color.RED);
        Rect bounds = new Rect(15, 320, 55, 500);
```

```java
            //设置图像显示的区域
            mShapeDrawable.setBounds(bounds);
            //绘制矩形图像
            mShapeDrawable.draw(canvas);
            /*##############椭圆图像 ##########*/
            /* 实例化ShapeDrawable对象并说明是绘制一个椭圆 */
            mShapeDrawable = new ShapeDrawable(new OvalShape());
            //得到画笔paint对象并设置其颜色
            mShapeDrawable.getPaint().setColor(Color.GREEN);
            //设置图像显示的区域
            mShapeDrawable.setBounds(70, 280, 150, 320);
            //绘制椭圆图像
            mShapeDrawable.draw(canvas);
            Path path1 = new Path();
            //设置多边形的点
            path1.moveTo(150+15, 120+80-50);
            path1.lineTo(150+80, 120+80-50);
            path1.lineTo(150+60, 120+140-50);
            path1.lineTo(150+20, 120+140-50);
            path1.lineTo(150+45, 120+140-70);
            //使这些点构成封闭的多边形
            path1.close();
            //PathShape后面两个参数分别是宽度和高度
            mShapeDrawable = new ShapeDrawable(new PathShape(path1,150,
150));
            mShapeDrawable.getPaint().setColor(Color.BLUE);
            //设置图像显示的区域
            mShapeDrawable.setBounds(100, 170, 200, 280);
            //绘制多边图像
            mShapeDrawable.draw(canvas);
        }
    }
```

Activity 文件 Paint_Color.Java 里的启动屏幕模式与前面介绍的有所不同,参看下面的主文件。

com/Paint_Color/Paint_Color.java 的主要内容如下:

```java
    package com.Paint_Color;
    import android.app.Activity;
    import android.os.Bundle;
    public class Paint_Color extends Activity {
        private PaintView mPaintView;
        @Override
        public void onCreate(Bundle savedInstanceState) {
            super.onCreate(savedInstanceState);
            mPaintView = new PaintView(this);
```

```
            setContentView(mPaintView);//启动绘图文件,不一样的地方
        }
    }
```
运行结果如图 6-5 所示。

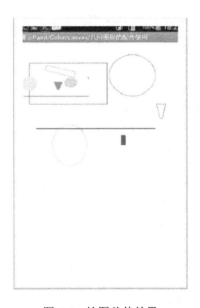

图 6-5　绘图总体效果

Canvas 类主要提供了三种构造方法,分别是构造一个空白的 Canvas 画布、在 Bitmap 位图文件中进行构造及从 GL 对象中进行构造,具体方法如下:

```
Canvas()            //构建一个空白画布,可以用 Bitmap 方法来绘制具体画布
Canvas(GL gl)       //这个方法在画 3D 图像时使用
Canvas(Bitmap bitmap)   //以 Bitmap 对象构建一个新画布,将所开发的内容描绘在此
画布上,注意 Bitmap 对象不能为空
```

此外,Canvas 类提供的应用方法中,有一些字段可以用来定义绘制方法,如 Canvas.HAS_ALPHA_LAYER_ SAVE_ FLAG 保存时需要 alpha 层。Canvas 类提供的方法很多,包括本身属性设置和相关图像操作方法等,大多数都是 Graphics 类绘图的常用方法。

6.2.5　绘制字符串

在实际开发中,有一些对话需要用动态的字符串进行提示,如游戏开发、图片浏览等,此时可以在 Canvas 画布中用绘制字符串的形式实现。对此,Android 系统提供了一系列 drawText 方法来绘制字符串。在绘制之前需要设置各种属性,包括设置画笔、画布和对象字符串的属性,然后用 FontMetrics 类来规划字体属性,再用 getFontMetrics 来获得系统字体的相关内容(完整源代码:AndroidDevelopment\Chapter6\Section6_4\ActivityStrings\ Activitystrings)。

先定义一个要显示的字符串:String string = " \n\nIf you shut door to all errors truth will be shut out."，然后重新定义 ondraw 方法：

```
public void onDraw(Canvas canvas){
    super.onDraw(canvas);
    //设置背景颜色
    canvas.drawColor(Color.DKGRAY);
    mPaint.setAntiAlias(true);
    if ( Count < 100 ){
        Count++;
    }
    mPaint.setColor(Color.WHITE);
    mPaint.setTextSize(25);
    canvas.drawText("Program running progress:"+ Count+"%......", 100, 370, mPaint);
    //绘制 stringsText：实现自动换行
    mTextstring.DrawText(canvas);
}
```

DrawText()方法的重新定义：

```
public void DrawText(Canvas canvas){
    for (int i = m_iCurLine, j = 0; i < m_iRealLine; i++, j++){
        if(j > m_ipageLineNum){
            break;
        }
        canvas.drawText((String)(m_String.elementAt(i)),m_iTextPosX,
        m_iTextPosY + m_iFontHeight * j, m_paint);
    }
}
```

运行结果如图 6-6 所示。

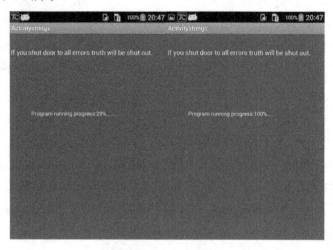

图 6-6　绘制字符串

6.2.6 android.graphics.Bitmap 类

该类属于位图操作类，Bitmap 为开发者提供了许多实用有效的方法，如：

Boolean compress（Bitmap.CompressFormat format,int quality,OutputStream stream）//压缩一个 Bitmap 对象，将与该对象相关的编码、画质像素保存到一个 OutputStream 文件中

创建位图对象 createBitmap 包含的 7 种方法如下：

```
//创建位图的各种方法
static Bitmap createBitmap(Bitmap src)
static Bitmap createBitmap(int[]colors,int width,int height, Bitmap.Config config)
static Bitmap reateBitmap(int[] colors, int offset, int stride, int width, int height,Bitmap.Config config)
static Bitmap createBitmap(Bitmap source, int x, int y, int width, int height, Matrix m, boolean filter)
static Bitmap createBitmap(int width, int height, Bitmap.Config config)
static Bitmap createBitmap(Bitmap source, int x, int y, int width, intheight)
static Bitmap createScaledBitmap(Bitmap src, int dstWidth, int dstHeight, boolean filter)
//创建一个可以缩放的位图对象
final int getHeight()    //获取高度
final int getWidth()     //获取宽度
final boolean hasAlpha()   //是否有透明通道
void setPixel(int x, int y, int color)    //设置某像素的颜色
int getPixel(int x, int y)   //获取某像素的颜色，Android 开发网提示这里返回的 int 型是 color 的定义[32]
```

1．图像绘制

Android 系统把图片等资源存在 res/drawable 里。Android 系统提供了 Bitmap 来存放这些资源，并显示在屏幕上，具体方法如下。先创建一个 Bitmap：

```
mBityun=((BitmapDrawable)getResources().getDrawable(R.drawable.yun)).getBitmap();
```

把资源图片与创建的 Bitmap 绑定，然后用 drawBitmap()方法把 mBityun 在画布上显示出来：

```
canvas.drawBitmap(mBityun, x, y, null); // 用 drawBitmap 方法绘制图像
```

完整源代码：AndroidDevelopment\Chapter6\Section6_5\ActivityBitmap\ActivityBitmap。com/ActivityBitmap/BitmapView.java 的主要内容如下：

```
public class BitmapView extends View implements Runnable{
    private Paint mPaint = null;
    Bitmap mBityueliang = null;
    Bitmap mBityun = null;
```

```java
        int miDTX = 0;
        public BitmapView(Context context){
            super(context);
            setmPaint(new Paint());
            miDTX = 0;
            mBityueliang = ((BitmapDrawable)
            getResources().getDrawable(R.drawable.yz)).getBitmap();
            mBityun = ((BitmapDrawable)
            getResources().getDrawable(R.drawable.yz1)).getBitmap();
            new Thread(this).start();
        }
        public void onDraw(Canvas canvas){
            super.onDraw(canvas);
            canvas.drawColor(Color.LTGRAY);
            BitmapView.drawImage(canvas, mBityueliang, 30, 10);
            BitmapView.drawImage(canvas, mBityun, miDTX+30,
            mBityueliang.getHeight()+20,mBityun.getWidth(),mBityun.getH
            eight
            (), 0, 0);
        }
```

com/ActivityBitmap/ActivityBitmap.java 的主要内容如下：

```java
    public class ActivityBitmap extends Activity{
        private BitmapView mBitmapView = null;
        /** Called when the activity is first created. */
        @Override
        public void onCreate(Bundle savedInstanceState){
            super.onCreate(savedInstanceState);
            mBitmapView = new BitmapView(this);
            setContentView(mBitmapView);
        }
        public boolean onKeyUp(int keyCode, KeyEvent event){
            super.onKeyUp(keyCode, event);
            return true;
        }
        public boolean onKeyDown(int keyCode, KeyEvent event){
            if ( mBitmapView == null ){
                return false;
            }
            return mBitmapView.onKeyDown(keyCode,event);
        }
    }
```

运行结果如图 6-7 所示。

图 6-7 绘制图片

2. 图像旋转

图像的旋转需要用到 set()和 reset()方法：

```
mMatrix.reset();        // 重置 mMatrix 对象
mMatrix.setRotate(Angle);//把角度变量与图像关联，然后改变角度实现旋转
```

用 createBitmap 方法按 mMatrix 的旋转属性创建一个新的 Bitmap 对象，再用 drawBitmap()方法把 Bitmap 图像显示出来（完整源代码：AndroidDevelopment\Chapter6\Section6_6\ActivityMatrix\ActivityMatrix）。

代码的主要内容如下：

```
public class MatrixView extends View implements Runnable{
    Bitmap mBityueliang = null;
    int Bityueliangwidth = 0;
    int Bityueliangheight = 0;
    floatAngle= 0.0f;
    Matrix mMatrix = new Matrix();
    public MatrixView(Context context){
        super(context);
        mBityueliang = ((BitmapDrawable) getResources().
        getDrawable(R.drawable.yz2)).getBitmap();
        Bityueliangwidth = mBityueliang.getWidth();
        Bityueliangheight = mBityueliang.getHeight();
        new Thread(this).start();
    }
    public void onDraw(Canvas canvas){
        super.onDraw(canvas);
        mMatrix.reset();
        canvas.drawColor(Color.LTGRAY);
```

```
            mMatrix.setRotate(Angle);
            Bitmap mBityueliang2 = Bitmap.createBitmap(mBityueliang, 0, 0,
            Bityueliangwidth,Bityueliangheight, mMatrix, true);
            MatrixView.drawImage(canvas, mBityueliang2,
            (320-Bityueliangwidth)/2,10);
            mBityueliang2 = null;
        }
        public boolean onTouchEvent(MotionEvent event){
            return true;
        }
        public boolean onKeyDown(int keyCode, KeyEvent event){
            if (keyCode == KeyEvent.KEYCODE_DPAD_LEFT){
                Angle--;
            }else if (keyCode == KeyEvent.KEYCODE_DPAD_RIGHT){
                Angle++;
            }
            return true;
        }
        public boolean onKeyUp(int keyCode, KeyEvent event){
            return false;
        }
        public boolean onKeyMultiple(int keyCode, int repeatCount, KeyEvent event){
            return true;
        }
        public void run(){
            while (!Thread.currentThread().isInterrupted()){
                try{
                    Thread.sleep(100);
                }catch (InterruptedException e){
                    Thread.currentThread().interrupt();
                }
                postInvalidate();
            }
        }
        public static void drawImage(Canvas canvas, Bitmap bitmap, int x, int y){
            canvas.drawBitmap(bitmap, x, y, null); // 绘制图像
        }
    }
```

com/ActivityMatrix/ActivityMatrix.java 的主要内容如下：

```
public class ActivityMatrix extends Activity{
    private MatrixView mMatrixView = null;
```

```
/* Called when the activity is first created. */
@Override
public void onCreate(Bundle savedInstanceState){
    super.onCreate(savedInstanceState);
    mMatrixView = new MatrixView(this);
    setContentView(mMatrixView);
}
public boolean onKeyUp(int keyCode, KeyEvent event){
    super.onKeyUp(keyCode, event);
    return true;
}
public boolean onKeyDown(int keyCode, KeyEvent event){
    if ( mMatrixView == null ){
        return false;
    }
    if ( keyCode ==  KeyEvent.KEYCODE_BACK){
        this.finish();
        return true;
    }
    return mMatrixView.onKeyDown(keyCode,event);
}
}
```

运行结果如图 6-8 所示。

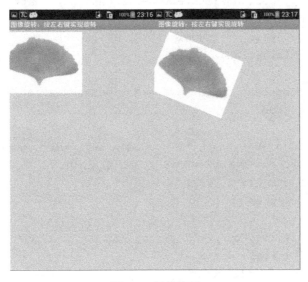

图 6-8 图像旋转

3. 图像缩放

图像缩放与旋转同样位于 Matrix 类中，其相关操作方法的特别之处是利用了 Matrix 中的 postScale 方法来设置图片缩放比例：mMatrix.postScale（float sx, float sy）（完整源代码：

AndroidDevelopment\Chapter6\Section6_7\ActivityPostScale\Activitypostscale）。
com/Activitypostscale/postScaleView.java 的主要内容如下：

```java
public class postScaleView extends View implements Runnable{
    Bitmap  mBityueliang= null;
    int Bityueliangwidth= 0;
    int Bityueliangheight= 0;
    float  Scale= 1.0f;
    Matrix mMatrix = new Matrix();
    public postScaleView(Context context){
        super(context);
        mBityueliang = ((BitmapDrawable) getResources().
        getDrawable(R.drawable.yz3)).getBitmap();
        Bityueliangwidth = mBityueliang.getWidth();
        Bityueliangheight = mBityueliang.getHeight();
        new Thread(this).start();
    }
    public void onDraw(Canvas canvas){
        super.onDraw(canvas);
        mMatrix.reset();
        mMatrix.postScale(Scale,Scale);
        canvas.drawColor(Color.LTGRAY);
        Bitmap mBityueliang2 = Bitmap.createBitmap(mBityueliang, 0, 0,
        Bityueliangwidth,Bityueliangheight, mMatrix, true);
        postScaleView.drawImage(canvas,mBityueliang2,
        (500-Bityueliangwidth)/2, 30);
        mBityueliang2 = null;
    }
    public boolean onTouchEvent(MotionEvent event){
        return true;
    }
    public boolean onKeyDown(int keyCode, KeyEvent event){
        if (keyCode == KeyEvent.KEYCODE_DPAD_UP){
            if ( Scale > 0.3 ){
                Scale-=0.1;
            }
        }else if (keyCode == KeyEvent.KEYCODE_DPAD_DOWN){
            if ( Scale < 1.5 ){
                Scale+=0.1;
            }
        }
        return true;
    }
    public boolean onKeyUp(int keyCode, KeyEvent event){
```

```
        return false;
    }
    public boolean onKeyMultiple(int keyCode, int repeatCount, KeyEvent
event){
        return true;
    }
    public void run(){
        while (!Thread.currentThread().isInterrupted()){
            try{
                Thread.sleep(120);
            }
            catch (InterruptedException e){
                Thread.currentThread().interrupt();
            }
            postInvalidate();
        }
    }
    public static void drawImage(Canvas canvas, Bitmap bitmap, int x, int y){
        canvas.drawBitmap(bitmap, x, y, null);
    }
}
```

运行结果如图 6-9 所示。

图 6-9 图像缩放

图像旋转与缩放的共同点主要是绘图原理：用 createBitmap 方法按 mMatrix 的旋转设置创建一个新的 Bitmap 位图对象，再把 Bitmap 图像用 drawBitmap() 方法显示出来。

常用图形的旋转、缩放等相关操作方法有：

```
void reset()    // 重置一个 matrix 对象
```

```
void set(Matrix src)  //复制一个源矩阵
boolean isIdentity()  //返回这个矩阵是否定义(已经有意义)
void setRotate(float degrees)  //指定一个角度以0、0为坐标进行旋转
void setRotate(float degrees, float px, float py)  //指定一个角度以px、py为坐标进行旋转
void setScale(float sx, float sy)  //缩放
void setScale(float sx, float sy, float px, float py)  //以px、py为坐标进行缩放
void setTranslate(float dx, float dy)  //平移
void setSkew(float kx, float ky, float px, float py)  //以px、py为坐标进行倾斜
void setSkew(float kx, float ky)  //倾斜[33]
```

4. 图像特效处理

图像特效处理主要是针对图像的像素进行处理，改变像素的组成属性就会有相应的效果，如透明度。Android 系统提供了 BitmapFactory 类来对像素进行操作。BitmapFactory 是 Bitmap 的操作类（制作工厂），常用的实现方法是用 getPixels()方法把图像的像素读取出来并存放在一个数组中，然后处理数组，最后用 setPixls()方法把处理过的像素数组设置到 Bitmap 对象中，即可实现特效。下面的例子是实现水波荡漾（完整源代码：Android Development\Chapter6\Section6_8\ActivityPixels\ActivityPixels）。

com/ActivityPixels/PixelsView.java 的主要内容如下：

```java
public class PixelsView extends View implements Runnable{
    int imageWIDTH;
    int imageHEIGHT;
    short[] bufer1;
    short[] bufer2;
    int[] Bitmap1;
    int[] Bitmap2;
    public PixelsView(Context context){
        super(context);
        Bitmap image = BitmapFactory.decodeResource(this.getResources(),
        R.drawable.yu);
    imageWIDTH = image.getWidth();
    imageHEIGHT = image.getHeight();
        bufer2 = new short[imageWIDTH * imageHEIGHT];
        bufer1 = new short[imageWIDTH * imageHEIGHT];
        Bitmap2 = new int[imageWIDTH * imageHEIGHT];
        Bitmap1 = new int[imageWIDTH * imageHEIGHT];
        image.getPixels(Bitmap1,0,imageWIDTH,0,0,imageWIDTH,
        imageHEIGHT);
        new Thread(this).start();
    }
```

```
void DropStone(int x, int y, int stonesize, int stoneweight) {
    for (int posx = x - stonesize; posx < x + stonesize; posx++)
        for (int posy = y - stonesize; posy < y + stonesize; posy++)
            if ((posx - x) * (posx - x) + (posy - y) * (posy - y)
                < stonesize * stonesize)
                    bufer1[imageWIDTH * posy + posx] = (short)
                    -stoneweight;
}
void RippleSpread(){
    for (int i = imageWIDTH; i < imageWIDTH * imageHEIGHT -
    imageWIDTH; i++){
            bufer2[i] = (short) (((bufer1[i - 1] + bufer1[i + 1] +
            bufer1[i - imageWIDTH] + bufer1[i + imageWIDTH]) >> 1)
            - bufer2[i]);
            bufer2[i] -= bufer2[i] >> 5;
    }
    short[] ptmp = bufer1;
    bufer1 = bufer2;
    bufer2 = ptmp;
}
void render(){
    int xoff, yoff;
    int k = imageWIDTH;
    for (int i = 1; i < imageHEIGHT - 1; i++){
            for (int j = 0; j < imageWIDTH; j++){
                    xoff = bufer1[k - 1] - bufer1[k + 1];
                    yoff=bufer1[k-imageWIDTH]-bufer1[k+imageWIDTH];
                    if ((i + yoff) < 0){
                            k++;
                            continue;
                    }
                    if ((i + yoff) > imageHEIGHT){
                            k++;
                            continue;
                    }
                    if ((j + xoff) < 0){
                            k++;
                            continue;
                    }
                    if ((j + xoff) > imageWIDTH){
                            k++;
                            continue;
                    }
```

```
                        int pos1, pos2;
                        pos1 = imageWIDTH * (i + yoff) + (j + xoff);
                        pos2 = imageWIDTH * i + j;
                        Bitmap2[pos2++] = Bitmap1[pos1++];
                        k++;
                    }
                }
            }
            public void onDraw(Canvas canvas){
                super.onDraw(canvas);
                canvas.drawColor(Color.LTGRAY);
                canvas.drawBitmap(Bitmap2, 0, imageWIDTH, 30, 60, imageWIDTH,
                                                imageHEIGHT, false, null);
            }
            public boolean onTouchEvent(MotionEvent event){
                return true;
            }
            public boolean onKeyDown(int keyCode, KeyEvent event){
                return true;
            }
            public boolean onKeyUp(int keyCode, KeyEvent event){
                DropStone(imageWIDTH/2, imageHEIGHT/2, 10, 30);
                return false;
            }
            public boolean onKeyMultiple(int keyCode, int repeatCount, KeyEvent
            event){
                return true;
            }
            public void run(){
                while (!Thread.currentThread().isInterrupted()){
                    try{
                        Thread.sleep(20);
                    }
                    catch (InterruptedException e){
                        Thread.currentThread().interrupt();
                    }
                    RippleSpread();
                    render();
                    postInvalidate();
                }
            }
        }
```

运行结果如图 6-10 所示。

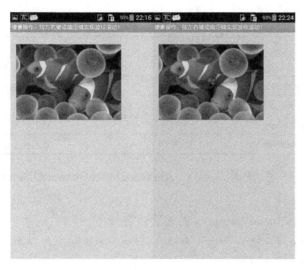

图 6-10 水波荡漾

BitmapFactory 类属于 Bitmap 对象的接口 I/O 类，为开发者提供了很多构造 Bitmap 对象的方法。例如，从某一个字节数组、文件、资源及输入流中创建一个 Bitmap 对象。下面为本类的全部成员，这些重载方法都很实用。

  static Bitmap decodeByteArray (byte[] data, int offset, int length) //从字节数组创建
  static Bitmap decodeByteArray (byte[] data, int offset, int length, BitmapFactory.Options opts)
  static Bitmap decodeFile (String pathName, BitmapFactory.Options opts) //从文件创建系统创建，注意路径要写全
  static Bitmap decodeFile (String pathName)
  static Bitmap decodeFileDescriptor (FileDescriptor fd, Rect outPadding, BitmapFactory.Options opts)　　//从输入流句柄创建
  static Bitmap decodeFileDescriptor (FileDescriptor fd)
  static Bitmap decodeResource (Resources res, int id)　　//从 Android 系统的 APK 文件资源中创建，提示是从/res/的 drawable 文件夹中创建
  static Bitmap decodeResource(Resources res, int id, BitmapFactory.Options opts)
  static Bitmap decodeResourceStream (Resources res, TypedValue value, InputStream is, Rect pad, BitmapFactory.Options opts)
  static Bitmap decodeStream (InputStream is)　　//从一个输入流中创建
  static Bitmap decodeStream (InputStream is, Rect outPadding, BitmapFactory.Options opts)

6.2.7　Shade 类

Android 系统中的 Shade 类主要用来对几何图像进行渲染，它有几个子类：BitmapShade、SweepGradient、LinearShade、RadialGradient、ComposeShade，如表 6-2 所示。

表 6-2 子类

类 型	主要用途
BitmapShade	用来渲染图像
SweepGradient	进行梯度渲染
LinearGradient	进行线性渲染
RadialGradient	进行环形渲染
ComposeShade	进行混合渲染

举例说明如下（完整源代码：AndroidDevelopment\Chapter6\Section6_9\Shade\ActivityShade。）

com/ActivityShade/ShadeView.java 的主要内容如下：

```java
public class ShadeView extends View implements Runnable{
    Bitmap  mBityun = null;
    int Bityunwidth = 0;
    int Bityunheight = 0;
    Paint  mPaint = null;
    Shader mBitmapShader = null;
    Shader mLinearGradient = null;
    Shader mComposeShader = null;
    Shader mRadialGradient = null;
    Shader mSweepGradient = null;
    ShapeDrawable mShapeDrawableyun = null;
    public ShadeView(Context context){
        super(context);
        mBityun = ((BitmapDrawable)
        getResources().getDrawable(R.drawable.sy3_1)).getBitmap();
        Bityunwidth = mBityun.getWidth();
        Bityunheight = mBityun.getHeight();
        mBitmapShader = new BitmapShader(mBityun,Shader.TileMode.REPEAT,
        Shader.TileMode.MIRROR);
        mLinearGradient = new LinearGradient(0,0,100,100,new
        int[]{Color.BLUE,Color.GREEN,Color.LTGRAY,Color.DKGRAY},null,
        Shader.TileMode.REPEAT);
        mComposeShader = new ComposeShader(mBitmapShader,mLinearGradient,
        PorterDuff.Mode.DARKEN);
        mRadialGradient = new RadialGradient(50,200,50,
        new int[]{Color.GRAY,Color.YELLOW,Color.BLUE,Color.DKGRAY},
        null,Shader.TileMode.REPEAT);
        mSweepGradient = new SweepGradient(30,30,new
        int[]{Color.BLUE,Color.TRANSPARENT,Color.CYAN,Color.WHITE},
        null);
        mPaint = new Paint();
        new Thread(this).start();
    }
```

```
public void onDraw(Canvas canvas){
    super.onDraw(canvas);
    canvas.drawColor(Color.LTGRAY);
    mShapeDrawableyun = new ShapeDrawable(new OvalShape());
    mShapeDrawableyun.getPaint().setShader(mBitmapShader);
    mShapeDrawableyun.setBounds(0,0, Bityunwidth, Bityunheight-90);
    mShapeDrawableyun.draw(canvas);
    mPaint.setShader(mLinearGradient);
    canvas.drawRect(Bityunwidth, 0, 320, 156, mPaint);
    mPaint.setShader(mComposeShader);
    canvas.drawRect(0, 300, Bityunwidth, 300+Bityunheight, mPaint);
    mPaint.setShader(mRadialGradient);
    canvas.drawCircle(80, 210, 60, mPaint);
    mPaint.setShader(mSweepGradient);
    canvas.drawRect(150, 160, 350, 250, mPaint);
    }
}
```
运行结果如图 6-11 所示。

图 6-11　图片渲染

6.3　动画设计

6.3.1　Tween 动画

Tween 动画通过对 View 的内容进行一系列的变换来实现，包括四种效果：移动动画、缩放比例动画、旋转动画、透明度渐变动画。

完整源代码：AndroidDevelopment\Chapter6\Section6_10\ ActivityTween\ActivityTween。com/ActivityTween/TweenView.java 的主要内容如下：

```
public class TweenView extends View{
```

```java
            private Animation mAnimationAlpha = null;
            private Animation mAnimationScale = null;
            private Animation mAnimationTranslate = null;
            private Animation mAnimationRotate = null;
            Bitmap Bityun = null;
            public TweenView(Context context){
                super(context);
                Bityun = ((BitmapDrawable)getResources().
                getDrawable(R.drawable.yz4)).getBitmap();
            }
            public void onDraw(Canvas canvas){
                super.onDraw(canvas);
                canvas.drawBitmap(Bityun, 30, 60, null);
            }
            public boolean onKeyUp(int keyCode, KeyEvent event){
                switch ( keyCode ){
                    case KeyEvent.KEYCODE_DPAD_LEFT:
                        mAnimationTranslate = new TranslateAnimation(20,150,20,
                        150);
                        mAnimationTranslate.setDuration(1200);
                        this.startAnimation(mAnimationTranslate);
                        break;
                    case KeyEvent.KEYCODE_DPAD_DOWN:
                        mAnimationScale =new ScaleAnimation(0.0f,2.0f,0.0f,2.0f,
                        Animation.RELATIVE_TO_PARENT, 0.5f,
                        Animation.RELATIVE_TO_PARENT, 0.5f);
                        mAnimationScale.setDuration(5000);
                        this.startAnimation(mAnimationScale);
                        break;
                    case KeyEvent.KEYCODE_DPAD_RIGHT:
                        mAnimationRotate=new RotateAnimation(0.0f, +360.0f,
                        Animation.RELATIVE_TO_SELF,0.5f,
                        Animation.RELATIVE_TO_SELF, 0.5f);
                        mAnimationRotate.setDuration(1200);
                        this.startAnimation(mAnimationRotate);
                        break;
                    case KeyEvent.KEYCODE_Q:
                        mAnimationAlpha = new AlphaAnimation(0.1f, 1.0f);
                        mAnimationAlpha.setDuration(4000);
                        this.startAnimation(mAnimationAlpha);
                        break;
                }
                return true;
            }
        }
```

1. Translate 位置移动动画、创建及属性设置

TranslateAnimation（float fromXDelta, float toXDelta, float fromYDelta, float toYDelta）；其中 fromXDelta 和 fromYDelta 为起始坐标；toXDelta 和 toYDelta 为终点坐标。

运行结果如图 6-12 所示。

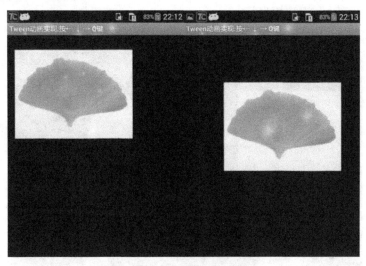

图 6-12　移动动画

2. Scale 缩放比例动画、创建及属性设置

ScaleAnimation（float fromX, float toX, float fromY, float toY, int pivotXType, float pivotXValue, int pivotYType, float pivotYValue）；其中 fromX 和 toX 分别表示 x 坐标轴上的缩放比例，fromY 和 toY 分别表示 y 坐标轴上的缩放比例，pivotXType 和 pivotYType 分别表示 x、y 的缩放模式，pivotXValue 和 pivotYValue 分别表示 x、y 坐标轴上的初始坐标。

运行结果如图 6-13 所示。

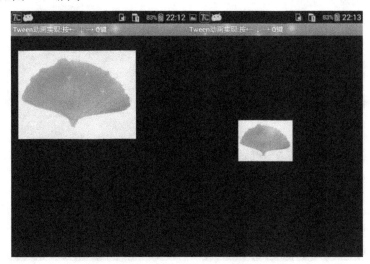

图 6-13　缩放比例动画

3. Rotate 旋转动画、创建及属性设置

RotateAnimation(float fromDegrees, float toDegrees, int pivotXType, float pivotXValue, int pivotYType, float pivotYValue);其中 fromDegrees 和 toDegrees 分别表示开始和结束时的角度,pivotXType 和 pivotYType 分别表示 x、y 坐标轴的旋转模式,pivotXValue 和 pivotYValue 分别表示 x、y 坐标轴上的初始坐标。

运行结果如图 6-14 所示。

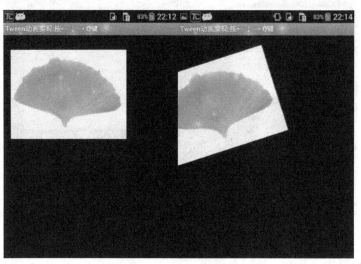

图 6-14　旋转动画

4. Alpha 透明度渐变动画、创建及属性设置

AlphaAnimation（float fromAlpha, float toAlpha）;其中 fromAlpha 和 toAlpha 分别表示起始和结束时的透明度,0.0 表示完全透明,1.0 表示完全不透明。

运行结果如图 6-15 所示。

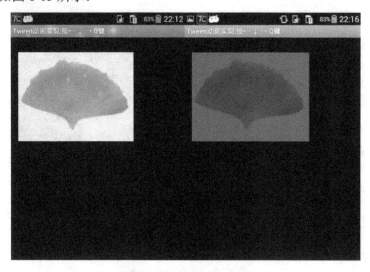

图 6-15　透明度渐变动画

6.3.2 Frame 动画

Frame 动画是最常见的动画之一，而且其实现思路很明了。首先创建一个 Drawable（如 myBitAnimation）对象来暂存图片资源，同时创建一个 AnimationDrawable 对象（如 frameAnimation）来表示动画。然后在每一个循环中通过 addFrame 方法逐个把每一帧（一张图片）加入 AnimationDrawable 对象里，最后用 Start 方法启动这个动画的播放。当然，播放后会用到 setOneShot 方法控制是否继续播放（完整源代码：AndroidDevelopment\Chapter6\Section6_11\ActivityFrame\ActivityFrame）。

com/ActivityFrame/FrameView.java 的主要内容如下：

```java
public class FrameView extends View{
    Context myContext = null;
    Drawable myBitAnimation = null;
    private AnimationDrawable frameAnimation = null;
    public FrameView(Context context){
        super(context);
        myContext = context;
        frameAnimation=newAnimationDrawable();
        for (int i = 0; i <=25; i++){
            int id = getResources().getIdentifier("p" + i, "drawable",
            myContext.getPackageName());
            myBitAnimation = getResources().getDrawable(id);
            frameAnimation.addFrame(myBitAnimation, 300);
        }
        frameAnimation.setOneShot( false );
        this.setBackgroundDrawable(frameAnimation);
    }
    public boolean onKeyUp(int keyCode, KeyEvent event){
        switch ( keyCode ){
            case KeyEvent.KEYCODE_DPAD_RIGHT:
            frameAnimation.start();
            break;
        }
        return true;
    }
}
```

运行结果如图 6-16 所示。

图 6-16 Frame 动画

第 7 章 Android 数据存储编程

Android 系统提供了多种方式让用户保存应用程序的数据，用户可以根据自己的需求选择相应的方法，如设置数据是否是应用程序私有的，数据是否能被其他程序访问，需要多少数据存储空间等。Android 系统中的数据存储方式包括文件存储、SQLite 数据库存储、SharedPreferences 存储、ContentProvider 存储、网络存储。

7.1 Android 中的文件操作

开发 Android 应用时总会与文件打交道，本节主要介绍常用到的文件操作。在 Android 系统中，一般都是通过调用 java.io.FileInputStream 和 java.io.File OutputStream 来实现对文件的读写的，java.io.File 则用于构造一个具体指向特定文件或文件夹的对象。每个应用程序都有一个存储自己私有数据的专门空间和具有访问 SD 卡的权限。而任何一个应用程序的私有数据其他程序没有权限访问，除非使用 ContentProvider 来提供数据共享。

7.1.1 File 类及常用方法

在 Android 系统中，操作一个文件（读写、创建文件或目录）是通过 File 类来完成的，这个操作和 Java 中完全一致。要操作一个文件，首先要构造文件对象，常用的构造文件对象的方法如下所示。

方式一：File(String pathname)

例如，File file = new File("/mnt/sdcard/test.txt")

方式二：File(String dir, String subpath)

例如，File file = new File("/mnt/sdcard/temp", "test.txt")

方式一中的输入参数为欲操作文件的绝对路径，方式二中的输入参数为欲操作文件所在的文件夹路径和文件名。

File 类的常用方法包括：

boolean exists() //测试文件是否存在
boolean delete() //删除此对象指定的文件
boolean createNewFile() //创建新的空文件
boolean isDirectory() //测试此对象表示的文件是否为目录
boolean mkdir() //创建由该对象表示的目录
boolean mkdirs() //创建包括父目录的目录
String getAbsolutePath() //返回此对象表示的文件的绝对路径名
String getName() //返回此对象表示的文件的名称
String getParent() //返回此对象路径名的上一级，若路径名没有上一级，则返回 null
String renameTo(newFile) //将文件名重命名为文件对象 newFile

File 类一般通过 FileInputStream 和 FileOutputStream 对文件进行操作。Context 提供了如下两个方法来打开本应用程序的数据文件 I/O 流。

```
FileInputStream openFileInput（String name）//打开应用程序的数据文件夹下的
```
name 文件对应输入流
```
FileOutputStream openFileOutput（String name, int mode）//打开应用程序的数
```
据文件夹下的 name 文件对应输出流。参数 mode 用于指定打开文件的模式（常用的打开文件的模式包括表 7-1 所示的几种情况）

表 7-1 文件操作方法

编号	常量	打开方式
1	MODE_APPEND	以在文件末尾写入数据方式打开文件
2	MODE_PRIVATE	以只有应用程序本身可对文件操作的方式打开或创建文件（可读可写）
3	MODE_WORLD_READABLE	以其他应用程序对文件有可读权限模式创建或打开文件
4	MODE_WORLD_WRITEABLE	以其他应用程序对文件有可写权限模式创建或打开文件

下面介绍创建文件、删除文件、重命名文件的代码示例。

```
public static final String FILE_NAME="myFile.txt"; //创建文件的名称，假
```
设文件名字为 "myFile.txt"

1. 创建文件示例

```
File file=new File(FileUtil.FILE_NAME);
 if(!file.exists())  //文件是否存在
    {
        try {
            file.createNewFile();//文件不存在，就创建一个新文件
            System.out.println("文件已经创建了");
        } catch (IOException e) {
            e.printStackTrace();
        }
    }
    else
    {
```

```
            System.out.println("文件已经存在");
            System.out.println("文件名："+file.getName());
            System.out.println("文件绝对路径为："+file.getAbsolutePath());
            System.out.println("文件相对路径为："+file.getPath());
            System.out.println("文件大小为："+file.length()+"字节");
            System.out.println("文件是否可读："+file.canRead());
            System.out.println("文件是否可写："+file.canWrite());
            System.out.println("文件是否隐藏："+file.isHidden());
        }
```

2. 删除文件示例

```
    File file=new File(FileUtil.FILE_NAME);
        //文件是否存在
        if(file.exists())
        {
            file.delete();
            System.out.println("文件已经被删除了");
        }
```

3. 为文件重命名示例

```
    File file=new File(FileUtil.FILE_NAME);
    File newFile=new File("anotherFile.txt");
    file.renameTo(newFile);
    System.out.println("文件已经成功地被命名了"+file.getName());
    //注意：当为文件重命名时，操作的仅仅是文件本身，内部的内容不会改变
```

7.1.2 文件 I/O

文件操作方式主要有：FileOutputStream()和FileInputStream()。

1. 从 I/O 流向文件中写入数据

```
    FileOutputStream(File file, Boolean append);  //当 append = ture 时写到文
    件的末尾，否则写到文件的开始
```

若要指定写入数据到文件的具体位置可使用write()方法：write(byte[] buffer,int offset,int count)，一般offset指定写入位置相对当前位置的偏移量，count指定写入数据的大小。一般使用File_name.write（buffer,offset,count）。示例代码如下：

```
    File_outPut = new FileOutputStream(file);
    String infoToWrite = "Android File I/O stream! I am the big man!";
    File_outPut.write(infoToWrite.getBytes());
    File_outPut.close();
```

2. 从文件向 I/O 流中写入数据

```
    File_inPut = new FileInputStream (file);
    int length = (int)file.length();  //获得文件的长度
    byte[] temp = new byte[length];
```

```
    File_inPut.read(temp, 0, length);//读入文件中的所有内容
    display = EncodingUtils.getString(temp,TEXT_ENCODING);
    File_inPut.close();//设置从文件读入I/O流中的数据以何种编码方式编码,以显示出来
完整源代码:AndroidDevelopment\Chapter7\Section7_1\FileOperate\FileOperate。
```

代码的主要内容如下:

```
    package com.wrd.fileoperate;
    import java.io.FileInputStream;
    import java.io.FileNotFoundException;
    import java.io.FileOutputStream;
    import java.io.PrintStream;
    import java.util.Scanner;
    import android.os.Bundle;
    import android.app.Activity;
    import android.view.Menu;
    import android.widget.TextView;
    import static java.lang.System.out;
    public class MainActivity extends Activity {
        private static final String FILENAME = "mldn.txt";
        TextView tv;
        protected void onCreate(Bundle savedInstanceState) {
            super.onCreate(savedInstanceState);
            setContentView(R.layout.activity_main);
            FileOutputStream output = null;
            try {
                output = openFileOutput(FILENAME, Activity.MODE_PRIVATE);
                out.println("ok");
            } catch (FileNotFoundException e) {
                e.printStackTrace();
            }
        }
    }
```

另外,要从工程中的 assets/文件夹下的文件中读出数据可使用 Context 对象提供的 getAssets().open(File_name)方法来实现,当然返回的是 InputStream 对象。

代码的主要内容如下:

```
    PrintStream out = new PrintStream(output);
    out.println("mmmmmmmmmm");
    out.println("ffff");
    out.close();
    tv = (TextView) findViewById(R.id.msg);
    FileInputStream input = null;
    try {
        input = openFileInput(FILENAME);
    } catch (FileNotFoundException e) {
```

```
        e.printStackTrace();
    }
    Scanner scanner = new Scanner(input);
    while (scanner.hasNext()) {
        tv.append(scanner.next() + "\n");
    }
    scanner.close();
public boolean onCreateOptionsMenu(Menu menu) {
    // Inflate the menu; this adds items to the action bar if it is present.
        getMenuInflater().inflate(R.menu.main, menu);
        return true;
    }
```

运行结果如图 7-1 所示。

图 7-1 I/O 流的操作

另外，如果 File 文件需要在程序中使用 SD 卡进行数据存储，需要在 AndroidMainfset.xml 文件中进行权限的配置，过程如下：

（1）SD 卡中创建与删除文件权限：

```
<uses-permission android:name=" android.permis sion.MOUNT_UNMOUNT_
FILESYSTEMS " />
```

（2）SD 卡中写入数据权限：

```
<uses-permission android:name=" android.permission.WRITE_EXTERNAL_
STORAGE " />
```

代码示意如下：

```
<?xml version="1.0" encoding="utf-8"?>
<manifest xmlns:android="http://schemas.android.com/apk/res/android"
        package="com.example.cxy.file">
<uses-permissionandroid:name="android.permission.MOUNT_FORMAT_FILESYSTEMS">
</uses-permission>
<uses-permission android:name="android.permission.WRITE_EXTERNAL_STORAGE">
</uses-permission>
```

```
<application
……
</application>

</manifest>
```

7.2 SharedPreferences

SharedPreferences 是 Android 平台上一个轻量级的存储类，主要是保存一些常用的配置，它提供了 Android 平台常规的 Long、Int、String 字符串型的保存类型。SharedPreferences 类似于 Windows 系统上的 ini 配置文件，但是它分为多种权限，可以全局共享访问。这是 Android 读写外部数据最简单的方法，以一种简单、透明的方式来保存用户个性化设置的字体、颜色、位置等参数信息。一般的应用程序都会提供类似于"设置"或"首选项"的界面，这些设置可以通过 Preferences 来保存，而不需要知道它到底以什么形式保存，保存在什么地方。在 Android 系统中，SharedPreferences 文件通常保存在/data/data/PACKAGE_NAME/shared_prefs 目录下。

如果应用程序需要保存相对较小的键值对集合，则可以使用 SharedPreferences API。SharedPreferences 对象用于指向包含了键值对的文件，并提供了一些读写这些键值对的方法。每个 SharedPreferences 文件均由框架管理，可以设为私有或共享。本节展示如何使用 SharedPreferences API 存储相对简单的数据。

7.2.1 获取 SharedPreferences 的句柄

有两种方法可以获得 SharedPreferences 对象，下面对这两种方法分别进行介绍。

1．getSharePreferences(parameter1, parameter2)

该方法中有两个参数，其中第一个参数 parameter1 表示需要访问的共享文件的名称，第二个参数 parameter2 表示该共享文件的权限。第二个参数 parameter2 有以下三种可能的取值：

MODE_PRIVATE：私有模式，只有创建该文件的程序可以访问该共享文件。
MODE_WORLD_READABLE：所有应用程序都可以读取该共享文件。
MODE_WORLD_WRITEABLE：所有应用程序都可以写入该共享文件。

2．getPreferences(parameter)

如果应用程序仅需要一个共享文件，则可以使用该方法。该方法有一个参数 parameter，用于表示共享文件的模式。因为只有一个共享文件，所以不需要提供共享文件的名称。

例如，以下代码在一个 Fragment 中执行，它访问了一个由资源字符串 R.string.preference_file_key 标识的文件，并且用私有模式打开它，因此这个文件只能由这个应用程序访问：

```
Context context = getActivity();
SharedPreferences sharedPref = context.getSharedPreferences(
    getString(R.string.preference_file_key),Context.MODE_PRIVATE);
```

命名共享参数文件时，应该使用唯一的、可识别的名称，如"com.example.myapp.PREFERENCE_FILE_KEY"。如果应用程序的 Activity 只有一个共享参数文件，可以使用 getPreferences()方法：

```
SharedPreferences sharedPref = getActivity().getPreferences(Context.MODE_PRIVATE);
```

7.2.2 写入共享文件

将数据写入共享文件需要四步。第一步，采用 getSharePreferences(parameter1, parameter2)方法或 getPreferences()方法获得 SharedPreferences 句柄；第二步，调用 SharedPreferences 类的 edit()方法创建 sharedPreferences.Editor 对象；第三步，调用诸如 putBoolean()、putString()、putInt()等方法写入具体的数据；第四步，调用 commit()方法提交新值。例如：

```
SharedPreferences sharedPref =
getActivity().getPreferences(Context.MODE_PRIVATE); //第一步，获得 SharedPre
-ferences 句柄
SharedPreferences.Editor editor = sharedPref.edit();// 第二步，创建 sharedPre
-ferences.Editor 对象
editor.putInt(getString(R.string.saved_high_score), newHighScore);// 第三
步，写入具体的数据
editor.commit();// 第四步，提交新值
```

7.2.3 读取共享文件

从共享文件中读取数据需要两步。第一步，采用 getSharePreferences（parameter1, parameter2）方法或 getPreferences()方法获得 SharedPreferences 句柄；第二步，调用诸如 getBoolean()、getString()、getInt()等方法读取具体的数据。例如：

```
SharedPreferences sharedPref = getActivity().getPreferences(Context.MODE_PRIVATE); //第一步，获得 SharedPreferences 句柄
int defaultValue = getResources().getInteger(R.string.saved_high_score_default);
long highScore = sharedPref.getInt(getString(R.string.saved_high_score), defaultValue); // 第二步，从共享文件中读取具体的数据
```

下面通过案例来演示如何使用 SharedPreferences。

MainActivity 的主要代码如下：

```
package com.itheima.rom;
import android.annotation.SuppressLint;
import android.app.Activity;
import android.content.SharedPreferences;
import android.content.SharedPreferences.Editor;
import android.os.Bundle;
import android.text.TextUtils;
import android.view.View;
```

```java
import android.widget.CheckBox;
import android.widget.EditText;
import android.widget.Toast;
public class MainActivity extends Activity {
    private EditText et_username;
    private EditText et_pwd;
    private CheckBox cb;
    //声明一个 SharedPreferences 对象
    private SharedPreferences sp;
    private static final String PWD = "123456";
    private static final String USERNAME = "wzy";
    @Override
    protected void onCreate(Bundle savedInstanceState) {
        super.onCreate(savedInstanceState);
        setContentView(R.layout.activity_main);
        sp = getSharedPreferences("info", MODE_PRIVATE);
        // 第二个参数表示该文件的访问模式为私有模式，这种模式比较安全
        et_username = (EditText) findViewById(R.id.et_username);
        et_pwd = (EditText) findViewById(R.id.et_pwd);
        cb = (CheckBox) findViewById(R.id.cb);

        String username = sp.getString("username", "");
        String pwd = sp.getString("pwd", "");
        et_username.setText(username);
        et_pwd.setText(pwd);
        /*
        * 从 sp 中获取用户信息，用户数据的回显
        * 第二个参数为默认返回值，即当要查找的 key-value 不存在时，返回的数据
        */

    }
    public void login(View view){
        String userName = et_username.getText().toString();
        String pwd = et_pwd.getText().toString();
        boolean checked = cb.isChecked();
        if (TextUtils.isEmpty(userName)) {
            Toast.makeText(this,"用户名不能为空！",Toast.LENGTH_SHORT).
            show();
            return ;
        }
        if (TextUtils.isEmpty(pwd)) {
            Toast.makeText(this,"密码不能为空！",Toast.LENGTH_SHORT).
            show();
```

```
                return ;
        }
        if (USERNAME.equals(userName)&&PWD.equals(pwd)) {
            if (checked) {
                Editor editor = sp.edit();
                editor.putString("username", userName);
                editor.putString("pwd", pwd);
                editor.commit();
                /*
                * 此处非常重要，执行完修改或写操作后只有调用 sp 的 commit
                方法，数据才会被保存下来
                */
            }else {//删除用户文件
                Editor editor = sp.edit();
                editor.clear();
                editor.commit();
            }
            Toast.makeText(this, "恭喜您，登录成功！", Toast.LENGTH_
            SHORT).show();
        }else {
            Toast.makeText(this, "对不起，登录失败！",
            Toast.LENGTH_SHORT).show();
        }
    }
}
```

运行上述程序后系统会自动创建一个文件：/data/data/com.itheima.rom/shared_prefs/info.xml。文件目录结构如图 7-2 所示。

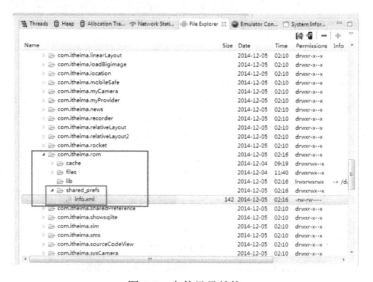

图 7-2　文件目录结构

7.3 SQLite 数据库数据存储

在 Android 应用程序中，通过 SQLiteDatabase 这个对象来完成连接 SQLite 数据库的操作。Android 中创建 SQLite 数据库的方法有以下两种。

1．用 SQLiteDatabase 的静态方式直接在 SQLite 中创建一个数据库

使用语句 SQLiteDatabase openDatabase（String path, SQLiteDatabase.CursorFactory Factory_name, int Flags）来打开位于 path 下的一个数据库，其中 Factory_name 为查询后的结果集存放的地方，也可以使用 Null，即用默认的 CursorFactory 对象作为查询结果集返回的存放地方。Flags 用于控制创建和打开的模式，其常用模式如表 7-2 所示。

表 7-2 Flags 的常用模式

标 号	常 量	功 能
1	OPEN_READONLY	以只读方式打开数据库
2	OPNE_READWRITE	以可写可读方式打开数据库
3	CREATE_IF_NECESSARY	当数据库不存在时就创建数据库，以可读可写方式打开
4	NO_LOCALIZED_COLLATORS	打开数据库，但是不根据本地语言对数据进行排序

2．调用 SQLiteDatabase 的 SQLiteDatabase openOrCreateDatabase（String path, SQLite Database.CurorFactory Factory_name）创建

该方法等同于 openDatabase（path, factory_name, CREATE_IF_NECESSARY）。与打开文件一样，同样可以使用 Context 对象调用 SQLiteDatabase openCreateDatabase（String path,Int mode, SQLiteDatabase.CurorFactory Factory_name）方法在应用程序的私有目录下创建或打开一个名为 SQLiteDB_name 的数据库（假设 Path=/data/data<pagename>/Database/SQLiteDB_name）。mode 的值一般是 MODE_PRIVATE、MODE_WORLD_READABLE、MODE_WORLD_WRITEABLE 等常量。开发者可以使用 databaseList()方法来获取私有数据库目录中的所有文件名，并返回一个存储有这些文件名的字符串数组。databaseDelete（String DB_name）方法用于删除 DB_name 数据库。

当用以上方法创建或打开数据库时，如果成功则返回 SQLiteDatabase 对象，否则会返回 Null 并抛出相对应的异常。无论是在何时何地用何种方法创建或者打开了一个数据库一定要记得在适当的时候关闭 close()方法，不然系统会抛出 IllegalStateExcpetion 异常。值得注意的是：Android 还支持在内存中创建数据库，只要使用 SQLiteDatabase create（SQLite Database.CursorFactort Factory_name）即可，创建成功则返回创建的 SQLiteDatabase 对象，否则是 Null（完整源代码：AndroidDevelopment\Chapter7\Section7_2\SqLite\sqlite）。

代码的主要内容如下：
```
package sqlite.activity;
```

```java
import sqlite.sqlite.SQLiteHelper;
import android.app.ListActivity;
import android.content.ContentValues;
import android.database.Cursor;
import android.database.sqlite.SQLiteDatabase;
import android.os.Bundle;

public class MailActivity extends ListActivity {
    /** Called when the activity is first created. */
    @Override
    public void onCreate(Bundle savedInstanceState) {
        super.onCreate(savedInstanceState);
        setContentView(R.layout.main);

        SQLiteHelper sh = new SQLiteHelper(this, "test_db");
        SQLiteDatabase db = sh.getReadableDatabase();
        db.execSQL("insert into user(id,name) values('1','x')");
        ContentValues values = new ContentValues();
        values.put("id", 2);
        values.put("name", "x1");
        db.insert("user", null, values);
        Cursor cursor = db.query("user", new String[]{"id","name"},
                "id=?", new String[]{"2"}, "", "", "");
        while(cursor.moveToNext()){
            System.out.println(cursor.getString(cursor.getColumnIndex
                                                            ("name")));
        }
        db.close();
    }
}
```

运行结果如图 7-3 所示。

SQLiteDatabase 提供了一系列对数据库进行操作的方法，如 execSQL()和 rawQuery()。除了直接对 SQL 语句分析的方法外，还有针对 INSERT、UPDATE、DELETE 和 SELECT 专门定义的方法，现总结如下：

（1）PublicvoidexecSQL（Stringsql）

PublicvoidexecSQL（Stringsql, Object[] bindArgs）执行一条非查询的 SQL 语句，不支持用";"隔开的多条 SQL 语句。其中参数 sql 为必须执行的 SQL 语句字符串。bindArgs 语句表达式中的"?"的占位参数只支持 String、byte[]、long、double 型数据。

（2）Public Curosr rawQuery（String sql,String[] args, public Cursor rawQueryWithFactory（SQLiteDatabase.CursorFactory Faxtory-

图 7-3 Android 创建数据库

Name, String sql, String[] args, String editTable））

执行一条 SQL 查询语句，并将查询的结果集封装成 Cursor 类对象返回。

args 语句表达式中的"?"的占位参数只支持 String 型参数。

FactoryName CursorFactory 对象用来构造查询完毕后返回结果集（Cursor）的子类。为 Null 时采用默认的构造。

editTable 为第一个可编辑的表名。

（3）Public long insert（String table,String nullColumnHack, ContentValues initialValues）

Public long insertOrThrow（String table, String nullColumnHack, ContentValues initial-Values）表示向指定的表中插入一行数据。

table：指定需要插入记录的表名。

nullColumnHack：需要插入表中的列名，一般不允许插入的所有列都为空。

initialValues：用来指明要插入型数据的 ContentValues 对象，即插入列名和相应的值。

（4）Public int update（String table, ContentValues values, String whereClause, String[] whereArgs）

更新表中指定行的数据。

table：指定要更新的表名。

values：表述更新后的类型数据（ContentValues 对象），即列名和列值的映射关系。

whereClause：可选的 where 语句，用以更新指定的行，当为 Null 时更新整个表，但注意该语句中不含有"WHERE"关键字。

（5）Public int delete（String table, String whereClause, String[] whereArgs）

删除表中的某行数据。

table：要删除行的表名。

whereClause：可选的过滤条件，即要从表中删除特定条件下的行。

whereArgs：where 语句中"?"的占位参数列表，只能用在 String 类型中。

（6）public Cursor query（String table，String[] columns，String selection，String[] selectionArgs，String groupBy String having, String orderBy, String limit）

（7）public Cursor query（Boolean distinct, String table，String[] columns, String selection，String[] selectionArgs，String groupBy String having, String orderBy, String limit）

（8）public Cursor queryWithFactory（SQLinteDatabase.CursorFactory cursorsFactory, Boolean distinct,String table，String[] columns，String selection，String[] selectionArgs, String groupBy String having, String orderBy）

（6）、（7）、（8）条语句均是按条件对表进行查询，并返回查询结果集以 Cursors 子类形式封装[34]。

table：查询表名。

selection：查询语句（SQL 语句）。

whereArgs where：语句中"?"的占位参数列表，只能用在 String 类型中。

groupBy:对查询结果集进行分组,是数据分组的依据。
having:根据设置条件过滤结果集,但需要配合 groupBy 语句使用。
orderBy:对结果集进行排序。
limit:对返回的行数进行限制,主要针对大型数据显示使用。
distinct:过滤结果集中重复出现的结果。
cursorFactory:返回结果集,此时结果集封装成 cursorFactory 形式。

第 3 篇

Android 高级编程

第 8 章　Android 多媒体编程

第 9 章　Android 网络与通信编程

第 10 章　AndroidOpenGL 应用开发

第 11 章　Android 传感器开发

第 12 章　Android NDK 开发技术

第 8 章

Android 多媒体编程

 ## 8.1 OpenCore 多媒体架构

Android 多媒体架构是基于第三方 PacketVideo 公司的 OpenCore 来实现的，支持通用的音频/视频/静态图像格式，包括 MPEG4、H.264、MP3、AAC、AMR、JPG、PNG、GIF 等。它按功能可分为两部分：一是音频/视频的回放（PlayBack）；二是音频/视频的记录（Recorder）。

OpenCore 的另外一个常用的名称是 PacketVideo，它是 Android 的多媒体核心。PacketVideo 是一家公司的名称，而 OpenCore 是这套多媒体框架的软件层的名称。对于 Android 的开发者而言，二者的含义基本相同。对比 Android 的其他程序库，OpenCore 的代码非常庞大，它是一个基于 C++的实现，定义了全功能的操作系统移植层，各种基本的功能均被封装成类的形式，各层次之间的接口多使用继承等方式。[35]OpenCore 是一个多媒体的框架，从宏观上来看，它主要包含以下两大方面的内容。

PVPlayer：提供媒体播放器的功能，完成各种音频（Audio）、视频（Video）流的回放（Playback）功能。

PVAuthor：提供媒体流记录的功能，完成各种音频（Audio）、视频（Video）流及静态图像的捕获功能。

PVPlayer 和 PVAuthor 以 SDK 的形式提供给开发者，可以在这个 SDK 之上构建多种应用程序和服务。在移动终端中常常使用的多媒体应用程序有媒体播放器、照相机、录像机、录音机等。

为了更好地组织整体的架构，OpenCore 在软件层分成以下几个层次（宏观层面）。

OSCL：Operating System Compatibility Library（操作系统兼容库），包含一些操作系统底层的操作，以便于更好地在不同操作系统中移植。它包含基本数据类型、配置、字符串工具、I/O、错误处理、线程等内容，类似于一个基础的 C++库。

PVMF：PacketVideo Multimedia Framework（PV 多媒体框架），在框架内实现一个文件的解析（Parser）和组成（Composer）、编解码的 NODE，也可以继承其通用的接口，在用户层实现一些 NODE。

PVPlayer Engine：PVPlayer 引擎。

PVAuthor Engine：PVAuthor 引擎。

事实上，OpenCore 中包含的内容非常多，从播放的角度，PVPlayer 的输入（Source）是文件或网络媒体流，输出（Sink）是音频/视频的输出设备，其基本功能包含媒体流控制、文件解析、音频/视频流的解码（Decode）等。除了从文件中播放媒体文件之外，它还包含与网络相关的 RTSP（Real Time Stream Protocol，实时流协议）流。在媒体流记录方面，PVAuthor 的输入（Source）是照相机、麦克风等设备，输出（Sink）是各种文件，包含流的同步、音频/视频流的编码（Encode）及文件的写入等功能。

使用 OpenCore 的 SDK 时，有可能需要在应用程序层实现一个适配器（Adaptor），然后在适配器之上实现具体的功能。PVMF 的 NODE 也可以基于通用的接口，在上层实现，并且以插件的形式使用。

针对第三方的多媒体编码/解码器、输入/输出设备等，OpenCore 有一套通用可扩展的接口，具体功能如下。

多媒体文件的播放、下载：包括 3GPP、MPEG-4、AAC、MP3 containers。

流媒体文件的下载、实时播放：包括 3GPP、HTTP、RTSP/RTP。

动态视频和静态图像的编码/解码：如 MPEG-4、H.263 及 AVC（H.264）、JPEG。

语音编码格式：AMR-NBandAMR-WB。

音乐编码格式：MP3、AAC、AAC+。

视频和图像格式：3GPP、MPEG-4、JPEG。

视频会议：基于 H324-Mstandard[35]。

在实际开发中开发者并不会去研究 OpenCore 的过程实现，因为 Android 系统提供了上层的 MediaAPI 给开发人员使用。可以看出，MediaPlayer 类提供了一个多媒体播放器的基本操作，包括播放、暂停、停止、设置音量等。Android 多媒体接口如表 8-1 所示。

表 8-1 Android 多媒体接口

文件	说明
Audio Manager.java	为上层应用提供了声音设置管理接口
AudioService.java	在 SystemServer 中启动，为所有的音频相关的设置提供服务。在 AudioService 类中定义了一个 AudioSystemThread 类，用来监控音频控制相关的信号，当有请求时，它会通过 AudioSystem 的接口实现音频的控制，这里的消息处理是异步的。此外，在 AudioService 中还抽象出了一套发送音频控制信号的接口，为 AudioManager 提供支持
AudioSystem.java	提供了音频系统的基本类型定义及基本操作的接口。它对应于 frameworks/base/core/jni/android_media_AudioTrack.cpp
Ringtone.java	为铃声、闹钟等提供了快速的播放接口
RingtoneManager.java	为铃声、闹钟等提供了快速的管理接口
AudioTrack.java	直接为 PCM 数据提供支持，对应于 frameworks/base/core/jni/android_media_AudioTrack.cpp
SoundPool.java	提供了播放声音的接口，在加载文件等方面做了优化
ToneGenneraor.java	提供了播放 DTMF tones 的支持，如电话的拨号音，对应于 frameworks/base/core/jni/android_media_toneGenerator.cpp
AudioRecord.java	这是音频系统对外的录制接口，对应于 frameworks/base/jni/android_media_AudioRecord.cpp

8.2 MediaPlayer 编程

Android 官方为开发者提供了 MediaPlayer 核心类,用来播放音乐。MediaPlayer 必须严格按照流程图进行操作,否则会出现错误。有些错误是底层向外抛出的,倘若严格按照状态图操作则一般不会出现问题。其状态流程图如图 8-1 所示。

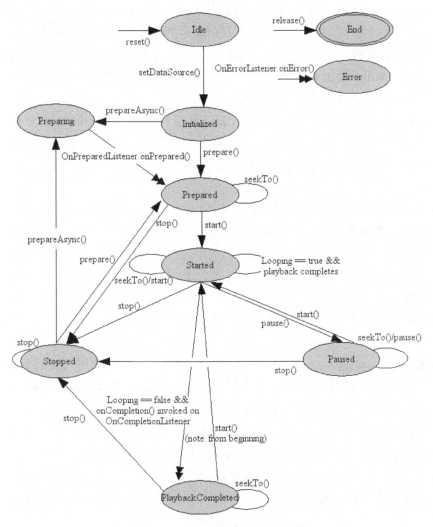

图 8-1　MediaPlayer 状态流程图

当有 MediaPlayer 对象被新建或调用 reset()方法之后,此时 MediaPlayer 处于 Idle 状态,只有在调用 release()方法后才处于 End 状态。当 MediaPlayer 对象不再被使用时,需要通过 release()来将其释放,使其处于 End 状态,以免造成不必要的错误和空间占用。当 MediaPlayer 对象处于 End 状态时,便不能再被使用,否则会触发 OnErrorListener.onError()事件。当

MediaPlayer 对象被新建时处于空闲状态，通过 creat()方法创建该对象之后处于准备状态。MediaPlayer 对象可以通过注册 setOnErrorListener（android.Media.MediaPlayer.OnErrorListener）方法来实现监控错误。如果发生了错误，MediaPlayer 对象就会处于错误状态，可以使用 reset()方法来恢复错误。在运行过程中，任何 MediaPlayer 对象都必须先处于准备状态，然后才开始播放。

MediaPlayer 对象有以下状态。

Idle 状态：当使用 new()方法创建一个 MediaPlayer 对象或调用了其 reset()方法时，该 MediaPlayer 对象处于 Idle 状态。调用 new()方法创建 MediaPlayer 对象进入 Idle 状态和调用 reset()方法进入 Idle 状态的一个重要差别就是：如果在这个状态下调用了 getDuration()等方法（相当于调用时机不正确），通过 reset()方法进入 Idle 状态时会触发 OnErrorListener.onError()，并且 MediaPlayer 会进入 Error 状态；如果是新创建的 MediaPlayer 对象，则不会触发 onError()，也不会进入 Error 状态。

End 状态：通过 release()方法可以进入 End 状态，只要 MediaPlayer 对象不再被使用，就应尽快将其通过 release()方法释放，以释放相关的软、硬件组件资源，这其中有些资源只有一份（相当于临界资源）。如果 MediaPlayer 对象进入 End 状态，则不会再进入任何其他状态。

Initialized 状态：这个状态比较简单，MediaPlayer 调用 setDataSource()方法就进入 Initialized 状态，表示此时要播放的文件已经设置好了。

Prepared 状态：初始化完成之后还需要调用 prepare()或 prepareAsync()方法，这两种方法中，一个是同步的，一个是异步的，只有进入 Prepared 状态，才表明 MediaPlayer 到目前为止都没有错误，可以进行文件播放。

Preparing 状态：这个状态比较好理解，主要是和 prepareAsync()配合，如果异步准备完成，会触发 OnPreparedListener.onPrepared()，进而进入 Prepared 状态。

Started 状态：显然，MediaPlayer 一旦准备好，就可以调用 start()方法了，这样 MediaPlayer 就处于 Started 状态，这表明 MediaPlayer 正处于播放文件的过程中。可以使用 isPlaying()测试 MediaPlayer 是否处于 Started 状态。如果播放完毕，而又设置了循环播放，则 MediaPlayer 仍然会处于 Started 状态，类似地，如果在该状态下 MediaPlayer 调用了 seekTo()或 start()方法，均可以让 MediaPlayer 停留在 Started 状态。

Paused 状态：Started 状态下，MediaPlayer 调用 pause()方法可以暂停 MediaPlayer，从而进入 Paused 状态，MediaPlayer 暂停后再次调用 start()则可以继续 MediaPlayer 的播放，转到 Started 状态。Paused 状态时可以调用 seekTo()方法，这是不会改变状态的。

Stopped 状态：Started 或 Paused 状态下均可调用 stop()停止 MediaPlayer，而处于 Stopped 状态的 MediaPlayer 要想重新播放，需要通过 prepareAsync()和 prepare()回到先前的 Prepared 状态重新开始才可以。

PlaybackCompleted 状态：文件正常播放完毕，没有设置循环播放就进入该状态，会触发 OnCompletionListener 的 onCompletion()方法。此时可以调用 start()方法重新从头播放文件；也可以调用 stop()方法停止 MediaPlayer，或者调用 seekTo()方法来重新定位播放位置。

Error 状态：如果由于某种原因 MediaPlayer 出现了错误，会触发 OnErrorListener.onError() 事件，此时 MediaPlayer 进入 Error 状态。及时捕捉并妥善处理这些错误是很重要的，可以帮助我们及时释放相关的软、硬件资源，也可以改善用户体验。通过 setOnErrorListener （android.media.MediaPlayer.OnErrorListener）可以设置该监听器。如果 MediaPlayer 进入 Error 状态，可以通过调用 reset()来恢复，使得 MediaPlayer 重新返回到 Idle 状态。[36]

MediaPlayer 类的常用方法如表 8-2 所示。

表 8-2 MediaPlayer 类的常用方法

方法	说明
MediaPlayer	构造方法
creat	创建一个要播放的多媒体
getCurrentPosition	得到当前播放的位置
getDuration	得到文件的时间
getVideoHeight	得到视频的高度
getVideoWidth	得到视频的宽度
isLooping	是否循环播放
isPlaying	是否正在播放
Pause	暂停
Prepare	准备（同步）
PerpareAsync	准备异步
Release	释放 MediaPlayer 对象
Reset	重置 MediaPlayer 对象
seekTo	制定播放的位置（以毫秒为单位的时间）
setAudioStreamType	设置流媒体的类型
setDataSource	设置多媒体数据来源
setDisplay	设置用 surfaceHolder 来显示
setLooping	设置是否循环播放
setOnBufferingUpdateLietener	网络流媒体的缓冲监听
setOnerrorListener	设置错误信息监听
setOnVideoSizeChangedListener	视频尺寸监听
setScreenOnWhilePlaying	设置是否使用 surfaceHolder 来显示
setVolume	设置音量
Start	开始播放
Stop	停止播放

8.2.1 MediaPlayer 主要接口定义

1．缓冲相关接口

接口名称：MediaPlayer.OnBufferingUpdateListener。

接口作用：定义一个回调接口，其作用是在流媒体缓冲状态发生改变时标明该状态。

接口方法：public abstract void onBufferingUpdate (MediaPlayer mp, int percent)。

方法作用：该方法在 MediaPlayer 通过 HTTP 下载缓冲视频流时回调，用以改变视频缓冲状态。

参数介绍：mp 为 MediaPlayer 实体对象；percent 为已经缓冲了或播放了的媒体流百分比。

2．播放完毕相关接口

接口名称：MediaPlayer.OnCompletionListener。

接口作用：在接口中定义了流媒体播放完毕后回调的方法。

接口方法：public abstract void onCompletion（MediaPlayer mp）。

方法作用：在媒体流播放完毕之后回调。

3．错误相关接口

接口名称：MediaPlayer.OnErrorListener。

接口作用：在该接口中定义回调方法，当在异步操作中出现错误时会回调该方法，其他情况下出现错误时直接抛出异常。

接口方法：public abstract boolean onError（MediaPlayer mp, int what, int extra）。

方法作用：异步操作中出现错误时回调该方法。

参数介绍：mp 为 MediaPlayer 实体类；what 为出现的错误类型，如 MEDIA_ERROR_UNKONWN（位置错误）、MEDIA_ERROR_SERVER_DIED（服务器错误）；extra 为针对于具体错误的附加码，用于定位错误的更详细信息，如 MEDIA_ERROR_IO（本地文件或网络相关错误）、MEDIA_ERROR_MALFORMAD（比特流不符合相关的编码标准和文件规范）、MEDIA_ERROR_UNSUPPORTED（框架不支持该功能）、MEDIA_ERROR_TIME_OUT（一些操作超时）。

方法执行结果：成功处理错误返回 True，处理失败返回 False，如果没有设定针对该方法的监听器，则直接调用 OnCompletionListener 监听器。

4．信息相关接口

接口名称：MediaPlayer.OnInfoListener。

接口作用：该接口定义了一个回调方法，该方法在媒体播放出现信息或警告时回调。

接口方法：public abstract boolean onInfo（MediaPlayer mp, int what, int extra）。

方法作用：出现信息或警告时回调。

参数介绍：mp 为 MediaPlayer 实体对象。what 为信息或警告的类型，如 MEDIA_INFO_UNKNOWN（未知的信息）、MEDIA_INFO_VEDIO_TRACK_LAGGING（视频过于复杂,解码太慢）、MEDIA_INFO_VEDIO_RENDERING_START（开始渲染第一帧）、MEDIA_INFO_BUFFRING_START（暂停播放开始缓冲更多数据）、MEDIA_INFO_BUFFERING_END（缓冲了足够的数据重新开始播放）、MEDIA_INFO_BAD_INTERLEAVING（错误交叉）、MEDIA_INFO_NOT_SEEKABLE（媒体不能够搜索）、

MEDIA_INFO_METADATA_UPDATE（更改媒体元数据）、MEDIA_INFO_UNSUPPORTED_SUBTITLE（不支持字幕）、MEDIA_INFO_SUBTITLE_TIMED_OUT（读取字幕使用时间过长）、extra 为信息或警告的类型。

返回值：如果处理了信息就会返回 True，没有处理则返回 False；如果没有注册该监听，就会忽略该信息。

5．准备播放相关接口

接口名称：MediaPlayer.OnPreparedListener。

接口作用：该接口中定义一个回调方法，该方法在进入 Prepared 状态并开始播放时回调。

接口方法：public abstract void onPrepared（MediaPlayer mp）。

方法作用：该方法在进入 Prepared 状态并开始播放时回调。

参数介绍：mp 为 MediaPlayer 实体对象。

6．查找操作相关接口

接口名称：MediaPlayer.OnSeekCompleteListener。

接口作用：该接口定义了一个回调方法，该方法在查找操作完成后回调。

接口方法：publicabstractvoidonSeekComplete（MediaPlayermp）。

方法作用：查找操作完成时回调该方法。

7．视频大小相关接口

接口名称：MediaPlayer.OnVideSizeChangedListener。

接口作用：该接口中定义了一个回调方法，当视频大小首次被知晓或更新时回调该方法。

接口方法：public abstract void onVideoSizeChanged（MediaPlayer mp, int width, int height）。

方法作用：视频大小更新时回调该方法，如果没有视频则返回 0。

参数介绍：width 为视频的宽度，height 为视频的高度。[37]

8.2.2　播放音乐实现

开发播放音乐的基本流程时，首先需要调用 reset()，重置 MediaPlayer 回到它的正常准备状态，然后通过所设置的路径找到歌曲并同时调用 prepare()方法和 start()方法（完整源代码：AndroidDevelopment\Chapter8\Section8_1\MusicPlayer\MusicPlayer）。

代码的主要内容如下：

```
package com.yarin.android.MusicPlayer;
import android.app.ListActivity;
import android.content.BroadcastReceiver;
import android.content.ComponentName;
import android.content.Context;
import android.content.Intent;
```

```java
import android.content.IntentFilter;
import android.content.ServiceConnection;
import android.database.Cursor;
import android.os.Bundle;
import android.os.IBinder;
import android.provider.MediaStore;
import android.view.View;
import android.widget.Button;
import android.widget.ListAdapter;
import android.widget.ListView;
import android.widget.SimpleCursorAdapter;
import android.widget.TextView;
public class MusicList extends ListActivity{
    private MusicPlayerService mMusicPlayerService = null;
    private MusicInfoController mMusicInfoController = null;
    private Cursor mCursor = null;
    private TextView mTextView = null;
    private Button mPlayPauseButton = null;
    private Button mStopButton = null;
    private ServiceConnection mPlaybackConnection=new
    ServiceConnection(){
      public void onServiceConnected(ComponentName className, IBinder
      service) {
          mMusicPlayerService = ((MusicPlayerService.LocalBinder)
          service). getService();
      }
      public void onServiceDisconnected(ComponentName className) {
            mMusicPlayerService = null;
      }
    };
    protected BroadcastReceiver mPlayerEvtReceiver = new BroadcastReceiver() {
      @Override
      public void onReceive(Context context, Intent intent) {
          String action = intent.getAction();
          if(action.equals(MusicPlayerService.PLAYER_PREPARE_END)){
              // will begin to play
              mTextView.setVisibility(View.INVISIBLE);
              mPlayPauseButton.setVisibility(View.VISIBLE);
              mStopButton.setVisibility(View.VISIBLE);
              mPlayPauseButton.setText(R.string.pause);
          } else if(action.equals(MusicPlayerService.PLAY_COMPLETED)){
              mPlayPauseButton.setText(R.string.play);
          }
```

```java
        }
    };
    public void onCreate(Bundle savedInstanceState) {
        super.onCreate(savedInstanceState);
        setContentView(R.layout.list_layout);
        MusicPlayerApp musicPlayerApp=(MusicPlayerApp)getApplication();
        mMusicInfoController = (musicPlayerApp).getMusicInfoController();
        // bind playback service
        startService(new Intent(this,MusicPlayerService.class));
        bindService(newIntent(this,MusicPlayerService.class),
        mPlaybackCon nection, Context.BIND_AUTO_CREATE);
        mTextView = (TextView)findViewById(R.id.show_text);
        mPlayPauseButton = (Button) findViewById(R.id.play_pause_btn);
        mStopButton = (Button) findViewById(R.id.stop_btn);
        mPlayPauseButton.setOnClickListener(new Button.
        OnClickListener(){
            public void onClick(View v) {
                // Perform action on click
                if (mMusicPlayerService != null && mMusicPlayerService.
                isPlaying()) {
                    mMusicPlayerService.pause();
                    mPlayPauseButton.setText(R.string.play);
                } else if (mMusicPlayerService != null){
                    mMusicPlayerService.start();
                    mPlayPauseButton.setText(R.string.pause);
                }
            }
        });
        mStopButton.setOnClickListener(new Button.OnClickListener() {
            public void onClick(View v) {
            // Perform action on click
            if (mMusicPlayerService != null ) {
                mTextView.setVisibility(View.VISIBLE);
                mPlayPauseButton.setVisibility(View.INVISIBLE);
                mStopButton.setVisibility(View.INVISIBLE);
                mMusicPlayerService.stop();
                }
            }
        });
        IntentFilter filter = new IntentFilter();
        filter.addAction(MusicPlayerService.PLAYER_PREPARE_END);
        filter.addAction(MusicPlayerService.PLAY_COMPLETED);
        registerReceiver(mPlayerEvtReceiver, filter);
```

```java
            }
            protected void onResume() {
                super.onResume();
                mCursor = mMusicInfoController.getAllSongs();
                ListAdapter adapter = new MusicListAdapter(this, android.R.lay
                out.simple_expandable_list_item_2, mCursor, new String[]{},
                new int[]{});
                setListAdapter(adapter);
            }
            protected void onListItemClick(ListView l, View v, int position, long id) {
                super.onListItemClick(l, v, position, id);
                if (mCursor == null ||mCursor.getCount() == 0) {
                    return;
                }
                mCursor.moveToPosition(position);
                String url = mCursor.getString(mCursor.getColumnIndexOrThrow
                (MediaStore.Audio.Media.DATA));
                mMusicPlayerService.setDataSource(url);
                mMusicPlayerService.start();
            }
        }
        /********************************
         *
         ********************************/
        class MusicListAdapter extends SimpleCursorAdapter {
            public MusicListAdapter(Context context, int layout, Cursor
            c,String[] from, int[] to) {
                super(context, layout, c, from, to);
            }
            public void bindView(View view, Context context, Cursor cursor) {
                super.bindView(view, context, cursor);
                TextView titleView = (TextView) view.findViewById(android.
                R.id.text1);
                TextView artistView = (TextView) view.findViewById(android.
                R.id.text2);
                titleView.setText(cursor.getString(cursor.getColumnIndexOr
                Throw(Me diaStore.Audio.Media.TITLE)));
                artistView.setText(cursor.getString(cursor.getColumnIndexOr
                Throw (MediaStore.Audio.Media.ARTIST)));
            }
            public static String makeTimeString(long milliSecs) {
                StringBuffer sb = new StringBuffer();
                long m = milliSecs / (60 * 1000);
```

```
            sb.append(m < 10 ? "0" + m : m);
            sb.append(":");
            long s = (milliSecs % (60 * 1000)) / 1000;
            sb.append(s < 10 ? "0" + s : s);
            return sb.toString();
        }
    }
```

音乐和程序是一起打包发布的。只需要将上例中的 setDatasource 方法替换，改成由 creat 方法来创建一个制定资源索引的 MediaPlayer 对象，其他操作基本一样，如下所示。

```
package com.yarin.android.MusicPlayer;
import java.io.IOException;
import android.app.Service;
import android.content.Intent;
import android.media.MediaPlayer;
import android.os.Binder;
import android.os.IBinder;
public class MusicPlayerService extends Service{
    private final IBinder mBinder = new LocalBinder();
    private MediaPlayer mMediaPlayer = null;
    public static final String PLAYER_PREPARE_END = "com.yarin.music
    playerservice.prepared";
    public static final String PLAY_COMPLETED = "com.yarin.music
    playerservice.playcompleted";
    MediaPlayer.OnCompletionListener mCompleteListener = new
    MediaPlayer.OnCompletionListener() {
        public void onCompletion(MediaPlayer mp) {
            broadcastEvent(PLAY_COMPLETED);
        }
    };
    MediaPlayer.OnPreparedListener mPrepareListener = new MediaPlayer.
    OnPreparedListener() {
        public void onPrepared(MediaPlayer mp) {
            broadcastEvent(PLAYER_PREPARE_END);
        }
    };
    private void broadcastEvent(String what){
        Intent i = new Intent(what);
        sendBroadcast(i);
    }
    public void onCreate(){
        super.onCreate();
        mMediaPlayer = new MediaPlayer();
        mMediaPlayer.setOnPreparedListener(mPrepareListener);
```

```java
            mMediaPlayer.setOnCompletionListener(mCompleteListener);
    }
    public class LocalBinder extends Binder{
        public MusicPlayerService getService(){
            return MusicPlayerService.this;
        }
    }
    public IBinder onBind(Intent intent){
        return mBinder;
    }
    public void setDataSource(String path){
        try{
            mMediaPlayer.reset();
            mMediaPlayer.setDataSource(path);
            mMediaPlayer.prepare();
        }catch (IOException e){
            return;
        }catch (IllegalArgumentException e){
            return;
        }
    }
    public void start(){
        mMediaPlayer.start();
    }
    public void stop(){
        mMediaPlayer.stop();
    }
    public void pause(){
        mMediaPlayer.pause();
    }
    public boolean isPlaying(){
        return mMediaPlayer.isPlaying();
    }
    public int getDuration(){
        return mMediaPlayer.getDuration();
    }
    public int getPosition(){
        return mMediaPlayer.getCurrentPosition();
    }
    public long seek(long whereto){
        mMediaPlayer.seekTo((int) whereto);
        return whereto;
    }
}
```

程序运行结果如图 8-2 所示。

creat 方法不仅可以加载应用程序中资源的索引，还可以加载一个 URL 地址，这就说明此时我们可以播放网络音乐。

音乐播放可以是一个 Service，Service 是不会直接与用户同时运行于后台的一类组件。当用户打开音乐播放后又进行别的操作时，音乐播放并不会停止，而是在后台继续播放，这就是播放音乐的 Service 进行控制的结果。Service 不能自己运行，需要通过某一个 Activity 或其他 Context 对象来调用。一个多媒体播放器的应用中会有许多 Activity，可以让用户选择歌曲并播放或暂停歌曲，但是音乐播放功能并没有所对应的 Activity。因此，音乐播放器可以使用 Context.startservice() 来启动一个 Service。

图 8-2　播放音乐视频

Service 的生命周期方法主要包括 onCreat()、onStart()和 onDestroy()。调用者（Serviceholder）通过 StartService()启动 Service 时会经历 onCreat()和 onStart()过程，Service 停止时直接进入销毁过程 onDestroy()。如果是调用者自己直接退出而没有调用 StopService()，Service 会一直在后台运行，直到下次调用者再启动并明确调用 StopService()为止。如果是 BindService 启动 Service，其只会运行 onCreat()，这时将调用者和 Service 绑定在一起，如果调用者退出，Service 就会调用 onUnbind()和 onDestroyed()，二者共存亡。Service 的 onCreat() 只调用一次，如果先是绑定（bind），则启动 Service 时就直接运行 Service 的 onStart()方法。如果先是启动 Service，则绑定时只运行 onBind()方法。如果先绑定，StopService()就不能用了，只能先执行 UnbindService()再执行 StopService()。[37]

8.2.3　播放视频实现

VideoView 类的常用方法如表 8-3 所示。

表 8-3　VideoView 类的常用方法

方　　法	说　　明
getBufferPercentage	得到缓冲的百分比
getCurrentPosition	得到当前播放的位置
getDuration	得到视频文件的时间
isPlaying	是否正在播放
pause	暂停
resolveAdjustedSize	调整视频显示大小
seekTo	指定播放位置
setMediaController	设置播放控制器模式（播放进度条）
setOnCompletionListener	当媒体文件播放完成时触发事件

(续表)

方法	说明
setOnErrorListener	错误监听
setVideoPath	设置视频源路径
setVideoURI	设置视频源地址
start	开始播放

播放视频的实现与播放音乐的实现很相似，读者可以自行去类比研究。

8.3 MediaRecoder 编程

MediaRecorder 类是 Android SDK 提供的一个专门用于音频/视频录制的工具，一般利用手机麦克风采集音频信息，利用摄像头采集图片信息。Mediarecorder 类的常用方法如表 8-4 所示。

表 8-4 Mediarecorder 类的常用方法

方法	说明
MediaRecorder	构造方法
getMaxAmplitude	得到目前为止最大的幅度
prepare	准备录音机
release	释放 MediaRecorder 对象
reset	重置 MediaRecorder 对象，使其为空闲状态
setAudioEncoder	设置音频编码
setAudioSource	设置音频源
setCamera	设置摄像机
setMaxDuration	设置最大期限
setMaxFileSize	设置文件的最大尺寸
setOnErrorListener	错误监听
setOutputFile	设置输出文件
setOutputFormat	设置输出文件的格式
setPreviewDisplay	设置预览
setViedeoEncoder	设置视频编码
setVideoFrameRate	设置视频帧的频率
setVideoSize	设置视频的宽度和高度（分辨率）
setVideoSource	设置视频源
start	开始录制
stop	停止录制

下面以录制 Media 音频资源为例，讲述 MediaRecorder 的使用。

（1）使用 new 创建一个实例 android.media.MediaRecorder；mr=new MediaRecorder()。

（2）创建一个 android.content.ContentValues 实例并设置一些标准的属性，如 TITLE、TIMESTAMP，最重要的是 MIME_TYPE；初始化 mr：mr.setAudioSource（MIC）/ setVideoSource，必须在配置 DataSource 之前调用。

（3）创建一个要放置文件的路径，可以通过 android.content.ContentResolver 在内容数据库中创建一个入口，并且自动地标记一个取得这个文件的路径。

（4）使用 MediaRecorder.setAudioSource()方法来设置音频资源；很可能会用到 MediaRecorder.AudioSource.MIC。

（5）使用 MediaRecorder.setOutputFormat()方法设置输出文件格式。

（6）使用 MediaRecorder.setAudioEncoder()方法来设置音频编码。

（7）最后，stop()和 release()在要结束时调用 prepare()和 start()所录制的音频。[38]

举例如下（完整源代码：AndroidDevelopment\Chapter8\Section8_2\MediaRecorderProject\MediaRecorderProject）：

```java
package org.lxh.demo;
import java.io.File;
import java.util.ArrayList;
import java.util.HashMap;
import java.util.List;
import java.util.Map;
import android.app.Activity;
import android.content.Intent;
import android.media.MediaRecorder;
import android.net.Uri;
import android.os.Bundle;
import android.os.Environment;
import android.view.View;
import android.view.View.OnClickListener;
import android.widget.AdapterView;
import android.widget.AdapterView.OnItemClickListener;
import android.widget.ImageButton;
import android.widget.ListView;
import android.widget.SimpleAdapter;
import android.widget.TextView;
public class MyMediaRecorderDemo extends Activity {
    private ImageButton record = null;
    private ImageButton stop = null;
    private TextView info = null;
    private ListView reclist = null;
    private SimpleAdapter recordSimpleAdapter = null;
    private MediaRecorder mediaRecorder = null;
    private boolean sdcardExists = false; //判断SD卡是否存在
    private File recordAudioSaveFileDir = null; //保存所有音频文件的文件夹
```

```java
        private File recordAudioSaveFile = null;// 每次保存音频文件的名称
        private String recordAudioSaveFileName = null;
        private String recDir = "mldnrec"; //保存的目录名称
        private boolean isRecord = false ; //录音的标志
        private List<Map<String,Object>> recordFiles = null ;
        @Override
        public void onCreate(Bundle savedInstanceState) {
            super.onCreate(savedInstanceState);
            super.setContentView(R.layout.main);
            this.record = (ImageButton) super.findViewById(R.id.record);
            this.stop = (ImageButton) super.findViewById(R.id.stop);
            this.info = (TextView) super.findViewById(R.id.info);
            this.reclist = (ListView) super.findViewById(R.id.reclist);
            // 如果存在则将状态给 sdcardExists 属性
            if ((this.sdcardExists = Environment.getExternalStorageState().
                equals(Environment.MEDIA_MOUNTED))) { //判断 SD 卡是否存在
                this.recordAudioSaveFileDir = new File(Environment.get
                External StorageDirectory().toString()+ File.separator+
                MyMediaRecorderDemo. this.recDir + File.separator);
                if (!this.recordAudioSaveFileDir.exists()) { //文件夹不存在
                    this.recordAudioSaveFileDir.mkdirs(); //创建文件夹
                }
            }
            this.stop.setEnabled(false) ;  //按钮现在不可用
            this.record.setOnClickListener(new RecordOnClickListenerImpl());
            this.stop.setOnClickListener(new StopOnClickListenerImpl());
            this.reclist.setOnItemClickListener(new OnItemClickListenerImpl
            ()) ;
            this.getRecordFiles() ;
        }
        private class RecordOnClickListenerImpl implements OnClickListener {
            @Override
            public void onClick(View v) {
                if(MyMediaRecorderDemo.this.sdcardExists) {//如果 SD 卡存在
                    MyMediaRecorderDemo.this.recordAudioSaveFile = new
                    File(MyMediaRecorderDemo.this.recordAudioSaveFileName);
                    MyMediaRecorderDemo.this.mediaRecorder = new Media
                    Recorder(); // 实例化对象
                    //在进行录制之前必须配置若干个参数

                    MyMediaRecorderDemo.this.mediaRecorder.setAudioSource
                    (MediaRecorder.AudioSource.MIC); //音频来源是 MIC
                    MyMediaRecorderDemo.this.mediaRecorder.setOutputFor
                    mat(MediaRecorder.OutputFormat.THREE_GPP);
```

```java
                MyMediaRecorderDemo.this.mediaRecorder.setAudioEncoder
                (MediaRecorder.AudioEncoder.DEFAULT);
                MyMediaRecorderDemo.this.mediaRecorder.setOutputFile
                (MyMediaRecorderDemo.this.recordAudioSaveFileName);
                try {    //进入就绪状态
                    MyMediaRecorderDemo.this.mediaRecorder.prepare();
                } catch (Exception e) {
                    //Log.i("MyMediaRecorderDemo", e.toString()) ;
                }
                MyMediaRecorderDemo.this.mediaRecorder.start() ;
                //开始录音
                MyMediaRecorderDemo.this.info.setText("正在录音中...") ;
                MyMediaRecorderDemo.this.stop.setEnabled(true);
                //停止录音按钮可以使用
                MyMediaRecorderDemo.this.record.setEnabled(false) ;
                MyMediaRecorderDemo.this.isRecord = true ;//正在录音
            }
        }
    }
    private class StopOnClickListenerImpl implements OnClickListener {
        @Override
        public void onClick(View v) {
            if(MyMediaRecorderDemo.this.isRecord) {  // 正在录音
                MyMediaRecorderDemo.this.mediaRecorder.stop() ;//停止
                MyMediaRecorderDemo.this.mediaRecorder.release() ;
                //释放资源
                MyMediaRecorderDemo.this.record.setEnabled(true) ;
                MyMediaRecorderDemo.this.stop.setEnabled(false) ;
                MyMediaRecorderDemo.this.info.setText("录音结束,文件路
                径为: "+MyMediaRecorderDemo.this.
                recordAudioSaveFileName);
                MyMediaRecorderDemo.this.getRecordFiles() ;
            }
        }
    }
    private void getRecordFiles(){ // 将一个文件夹之中的全部文件列出
        this.recordFiles = new ArrayList<Map<String, Object>>();
        if(this.sdcardExists) {    //有 SD 卡存在
            File files [] = this.recordAudioSaveFileDir.listFiles() ;
            // 列出目录中的文件
            for (int x = 0; x < files.length; x++) {
                Map<String, Object> fileInfo = new HashMap<String,
                Object>();
                fileInfo.put("filename", files[x].getName()) ;
```

```java
                    this.recordFiles.add(fileInfo) ;
                }
                this.recordSimpleAdapter = new SimpleAdapter(this,this.
                record Files, R.layout.recordfiles,new String[]
                { "filename" }, new int[] { R.id. filename });
                this.reclist.setAdapter(this.recordSimpleAdapter) ;
            }
        }
        private class OnItemClickListenerImpl implements OnItemClickListener {
            @Override
            public void onItemClick(AdapterView<?> parent, View view, int
            position,long id) {
                if(MyMediaRecorderDemo.this.recordSimpleAdapter.getItem
                (position)instanceof Map) {
                    Map<?, ?> map =(Map<?, ?>)MyMediaRecorderDemo.this.
                    recordSimpleAdapter.getItem(position);
                    Uri uri = Uri.fromFile(new File(MyMediaRecorderDe mo.
                    this.recordAudioSaveFileDir.toString()+File.
                    separator+map.get("filename")));
                    Intent intent = new Intent(Intent.ACTION_VIEW) ;
                    intent.addFlags(Intent.FLAG_ACTIVITY_NEW_TASK) ;
                    intent.setDataAndTy pe(uri, "audio/mp3") ;
                    MyMediaRecorderDemo. this.startActivity(intent) ;
                }
            }
        }
    }
```

程序运行结果如图 8-3 所示。

图 8-3 录制音频实现

8.4 Camera 编程

通常情况下，Camera UI 逻辑设计主要实现以下功能。

（1）拍照功能：在 Camera 模块被启动之后，首先进入拍照预览界面，单击预览界面的某处进行拍照，拍完一张照片之后，会再次回到拍照预览界面，所拍照片会被自动存储。

（2）拍照设置：在拍照预览模块下，可以设置菜单选项并进行拍照详细设置、拍照模式的切换和变焦切换。

（3）摄像预览：在拍照预览模块下，通过单击选项菜单可以进入摄像预览模块，在预览界面按下确认键可以开始摄像，在摄像模式下可以执行暂停或停止操作，摄像暂停/停止后，应回到摄像预览模式。

（4）视频回放：在视频预览模式下，通过一些操作或设置可进入视频回放模式。需要在屏幕显示时选中首帧进行回放，在回放的过程中用户能够执行暂停或停止等操作，执行暂停操作后可停留在视频播放某一帧所在界面，执行停止操作后可回到视频文件的浏览界面。

Camera 应用层设计主要实现以下模块的功能。

（1）Camera 应用接口模块，其功能相当于主控模块。

（2）拍照相关模块：实现有关拍照的功能。

（3）视频相关模块：实现有关摄像的功能。

（4）选项相关功能：实现照片属性设定功能。

（5）省电模块：实现待机功能。

所有设置完毕，用户可以通过按确认键直接拍照。[39]

Camera 是一个专门用来连接和断开相机服务的类，它包括如下几个事件。

Camera.AutoFocusCallback：自动调焦功能。

Camera.ErrorCallback：错误信息捕捉。

Camera.Parameters：相机的属性参数。

Camera.PictureCallback：拍照产生图片时触发。

Camera.previewCallback：相机预览设置。

Camera.ShutterCallback：快门设置。

Camera.Size：图片的尺寸。

Camera 没有构造方法，可以直接通过 open() 方法来打开相机设备，然后通过 Camera.Parameters 对相机的一些属性进行设置，如图片的格式大小等。

Takepicture 要实现三个回调函数，分别是 Camera.ShutterCallback 和两个 Camera.PictureCallback。拍照之后要取得图像数据就需要实现 Camera.PictureCallback 的 OnPictureTaken 方法。可以使用 SurfaceView 类来预览拍摄的效果。

Camera 类的方法如表 8-5 所示。

表 8-5 Camera 类的方法

方 法	说 明
autofucus	设置自动对焦
getParameters	得到相机的参数
open	启动相机服务
release	释放相机服务
setParameters	设置预览的参数
setPreviewDisplay	设置预览
startPreview	开始预览
stopPreview	停止预览
takePicture	拍照

举例说明（完整源代码：AndroidDevelopment\Chapter8\Section8_3\AndroidCamera\Android-Camera）：

```java
package com.example.cam;
import java.io.File;
import java.io.FileNotFoundException;
import java.io.FileOutputStream;
import java.io.IOException;
import android.app.Activity;
import android.content.Context;
import android.content.Intent;
import android.hardware.Camera;
import android.hardware.Camera.PictureCallback;
import android.hardware.Camera.ShutterCallback;
import android.net.Uri;
import android.os.AsyncTask;
import android.os.Bundle;
import android.os.Environment;
import android.util.Log;
import android.view.SurfaceView;
import android.view.View;
import android.view.View.OnClickListener;
import android.view.View.OnLongClickListener;
import android.view.ViewGroup.LayoutParams;
import android.view.Window;
import android.view.WindowManager;
import android.widget.Button;
import android.widget.FrameLayout;
import android.widget.Toast;
public class CamTestActivity extends Activity {
```

```java
private static final String TAG = "CamTestActivity";
Preview preview;
Button buttonClick;
Camera camera;
Activity act;
Context ctx;
@Override
public void onCreate(Bundle savedInstanceState) {
    super.onCreate(savedInstanceState);
    ctx = this;
    act = this;
    requestWindowFeature(Window.FEATURE_NO_TITLE);
    getWindow().addFlags(WindowManager.LayoutParams.FLAG_FULL
    SCREEN);
    setContentView(R.layout.main);
    preview = new Preview(this, (SurfaceView)findViewById(R.id.
    surface View));
    preview.setLayoutParams(newLayoutParams(LayoutParams.MATCH_
    PARENT,LayoutParams.MATCH_PARENT));
    ((FrameLayout) findViewById(R.id.layout)).addView(preview);
    preview.setKeepScreenOn(true);
    preview.setOnClickListener(new OnClickListener() {
        @Override
        public void onClick(View arg0) {
            camera.takePicture(shutterCallback,rawCallback,
            jpegCallback);
        }
    });
    Toast.makeText(ctx, getString(R.string.take_photo_help), Toast.
    LENG TH_LONG).show();
}
@Override
protected void onResume() {
    super.onResume();
    int numCams = Camera.getNumberOfCameras();
    if(numCams > 0){
        try{
            camera = Camera.open(0);
            camera.startPreview();
            preview.setCamera(camera);
        } catch (RuntimeException ex){
            Toast.makeText(ctx, getString(R.string.camera_not_found),
            Toast.LENGTH_LONG).show();
```

```java
                }
            }
        }
        @Override
        protected void onPause() {
            if(camera != null) {
                camera.stopPreview();
                preview.setCamera(null);
                camera.release();
                camera = null;
            }
            super.onPause();
        }
        private void resetCam() {
            camera.startPreview();
            preview.setCamera(camera);
        }
        private void refreshGallery(File file) {
            Intent mediaScanIntent = new Intent( Intent.ACTION_MEDIA_SCANNER_SCAN_FILE);
            mediaScanIntent.setData(Uri.fromFile(file));
            sendBroadcast(mediaScanIntent);
        }
        ShutterCallback shutterCallback = new ShutterCallback() {
            public void onShutter() {
                //Log.d(TAG, "onShutter'd");
            }
        };
        PictureCallback rawCallback = new PictureCallback() {
            public void onPictureTaken(byte[] data, Camera camera) {
                //Log.d(TAG, "onPictureTaken - raw");
            }
        };
        PictureCallback jpegCallback = new PictureCallback() {
            public void onPictureTaken(byte[] data, Camera camera) {
                new SaveImageTask().execute(data);
                resetCam();
                Log.d(TAG, "onPictureTaken - jpeg");
            }
        };
        private class SaveImageTask extends AsyncTask<byte[], Void, Void> {
            @Override
            protected Void doInBackground(byte[]... data) {
```

```
        FileOutputStream outStream = null;
        try {
            File sdCard = Environment.getExternalStorageDirectory();
            File dir = new File (sdCard.getAbsolutePath() +
            "/camtest");
            dir.mkdirs();
            String fileName = String.format("%d.jpg", System.
            current TimeMillis());
            File outFile = new File(dir, fileName);
            outStream = new FileOutputStream(outFile);
            outStream.write(data[0]);
            outStream.flush();
            outStream.close();
            Log.d(TAG,"onPictureTaken-wrote bytes: + data.length
            +" to " + outFile.getAbsolutePath());
            refreshGallery(outFile);
        } catch (FileNotFoundException e) {
            e.printStackTrace();
        } catch (IOException e) {
            e.printStackTrace();
        } finally {
        }
        return null;
    }
  }
}
```

程序运行结果如图 8-4 所示。

图 8-4　拍照实现

第 9 章 Android 网络与通信编程

网络编程是任何一个 Android 程序员必备的技能。Android 常用的有三种网络通信方式，即 Socket、HttpClient、HttpURLConnection。网络编程必须添加权限：<uses-permission android:name="android.permission.INTERNET" />。本章将介绍 HTTP 协议原理、Android 网络编程、WiFi 编程、蓝牙编程等内容。

9.1 HTTP 协议原理

TCP/IP 协议的七层协议模型如表 9-1 所示。

表 9-1　七层协议模型

OSI 中的层	功　　能	TCP/IP 协议族
应用层	文件传输、电子邮件、文件服务、虚拟终端	TFTP、HTTP、SNMP、FTP、SMTP、DNS、RIP、Telnet
表示层	数据格式化、代码转换、数据加密	没有协议
会话层	解除或建立与其他节点的联系	没有协议
传输层	提供端对端的接口	TCP、UDP
网络层	为数据包选择路由	IP、ICMP、OSPF、BGP、IGMP、ARP、RARP
数据链路层	传输有地址的帧及错误检测功能	SLIP、CSLIP、PPP、MTU
物理层	以二进制数据形式在物理媒体上传输数据	ISO2110、IEEE802、IEEE802.2

9.1.1　HTTP 简介

HTTP 是一个属于应用层的面向对象的协议，由于其简捷、快速的方式，适用于分布式超媒体信息系统。它于 1990 年提出，经过几年的使用与发展，得到了不断的完善和扩展。HTTP 的主要特点如下。

（1）支持 C/S（客户/服务器）模式。

（2）简单快速：客户向服务器请求服务时，只需要传送请求方法和路径。请求方法常用的有 GET、HEAD、POST，每种方法规定了客户端与服务器端不同的联系类型。由于 HTTP 协议简单，使得 HTTP 服务器的程序规模小，因而通信速度很快。

（3）灵活：HTTP 允许传输任意类型的数据对象。正在传输的类型由 Content-Type 加以标记。

（4）无连接：无连接的含义是限制每次连接只处理一个请求。服务器处理完客户的请求并收到客户的应答后，即断开连接。采用这种方式可以节省传输时间。

（5）无状态：HTTP 是无状态协议，无状态是指协议对于事务处理没有记忆能力。无状态意味着如果后续处理需要前面的信息，则必须重传，这样可能导致每次连接传送的数据量增大。另外，当服务器不需要先前信息时它的应答就较快。

HTTP URL 的格式如下：

$$http://host[":"port][abs\_path]$$

其中，http 表示要通过 HTTP 来定位网络资源；host 表示合法的 Internet 主机域名或 IP 地址；port 指定一个端口号，若为空则使用默认端口 80；abs_path 指定请求资源的 URI（Web 上任意的可用资源）。

HTTP 有两种报文，分别是请求报文和响应报文，下面先来了解请求报文。

9.1.2　HTTP 的请求报文

通常来说一个 HTTP 请求报文由请求行、请求报头、空行和请求数据 4 部分组成。

1．请求行

请求行由请求方法、URL 字段和 HTTP 协议版本组成，格式如下：

$$Method\ Request\text{-}URI\ HTTP\text{-}Version\ CRLF$$

其中，Method 表示请求方法；Request-URI 是一个统一资源标识符；HTTP-Version 表示请求的 HTTP 协议版本；CRLF 表示回车和换行（除了作为结尾的 CRLF 外，不允许出现单独的 CR 或 LF 字符）。

HTTP 请求方法有 8 种，分别是 GET、POST、DELETE、PUT、HEAD、TRACE、CONNECT、OPTIONS。其中 PUT、DELETE、POST、GET 分别对应改、删、增、查。移动开发最常用的是 POST 和 GET。

（1）GET：请求获取 Request-URI 所标识的资源。

（2）POST：在 Request-URI 所标识的资源后附加新的数据。

（3）HEAD：请求获取由 Request-URI 所标识的资源的响应消息报头。

（4）PUT：请求服务器存储一个资源，并用 Request-URI 作为其标识。

（5）DELETE：请求服务器删除 Request-URI 所标识的资源。

（6）RACE：请求服务器回送收到的请求信息，主要用于测试或诊断。

（7）CONNECT：HTTP/1.1 协议中预留给能够将连接改为管道方式的代理服务器。

（8）OPTIONS：请求查询服务器的性能，或者查询与资源相关的选项和需求。

例如，访问 CSDN 博客地址的请求行是：

GET http://blog.csdn.net/itachi85 HTTP/1.1

2．请求报头

在请求行之后会有 0 个或多个请求报头，每个请求报头都包含一个名字和一个值，它们之间用"："分隔。请求报头会以一个空行发送回车符和换行符，通知服务器后面不再请求报头信息了。关于请求报头，会在后面的消息报头一节（9.1.4 节）给出统一的解释。

3．请求数据

请求数据不在 GET 方法中使用，而是在 POST 方法中使用。POST 方法适用于需要客户填写表单的场合。与请求数据相关的最常用的请求报头是 Content-Type 和 Content-Length。

9.1.3　HTTP 的响应报文

HTTP 响应报文的一般格式如图 9-1 所示。

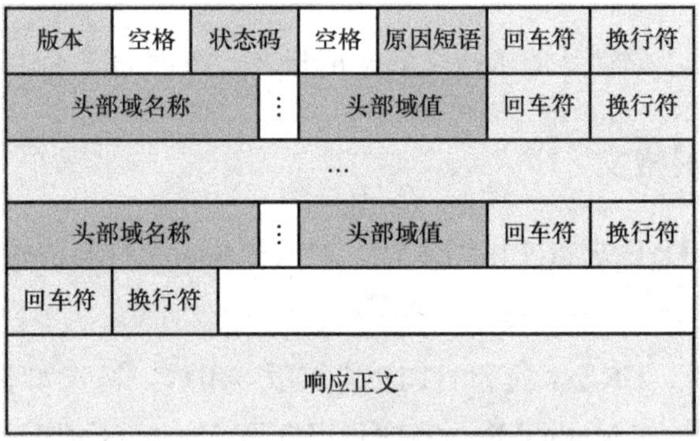

版本	空格	状态码	空格	原因短语	回车符	换行符
头部域名称	:	头部域值		回车符	换行符	
...						
头部域名称	:	头部域值		回车符	换行符	
回车符	换行符					
响应正文						

图 9-1　HTTP 响应报文的一般格式

HTTP 的响应报文由状态行、消息报头、空行、响应正文组成。消息报头后面会讲到，响应正文是服务器返回资源的内容。下面先来看看状态行。

状态行格式如下：

HTTP-Version Status-Code Reason-Phrase CRLF

其中，HTTP-Version 表示服务器 HTTP 协议版本；Status-Code 表示服务器发回的响应状态代码；Reason-Phrase 表示状态代码的文本描述。状态代码由三位数字组成，第一个数字定义了响应的类别，且有五种可能的取值，如表 9-2 所示。

表 9-2　状态代码的取值

100~199	指示信息，表示请求已接收，继续处理
200~299	请求成功，表示请求已被成功接收、理解、接收
300~399	重定向，要完成请求必须进行更进一步的操作
400~499	客户端错误，请求有语法错误或请求无法实现
500~599	服务器端错误，服务器未能实现合法的请求

常见的状态代码如下。

（1）200 OK：客户端请求成功。

（2）400 Bad Request：客户端请求有语法错误，不能被服务器所理解。

（3）401 Unauthorized：请求未经授权，这个状态代码必须和 WWW-Authenticate 报头域一起使用。

（4）403 Forbidden：服务器收到请求，但是拒绝提供服务。

（5）500 Internal Server Error：服务器发生不可预期的错误。

（6）503 Server Unavailable：服务器当前不能处理客户端的请求，一段时间后可能恢复正常。

例如，访问某 CSDN 博客地址响应的状态行是：

HTTP/1.1

200 OK

9.1.4 HTTP 的消息报头

消息报头分为通用报头、请求报头、响应报头、实体报头等。消息报头由键值对组成，每行一对，关键字和值用英文冒号"："分隔。

1．通用报头

通用报头既可以出现在请求报头中，也可以出现在响应报头中。

Date：表示消息产生的日期和时间。

Connection：允许发送指定连接的选项，如果指定连接是连续的或者指定"close"选项，则通知服务器在响应完成后关闭连接。

Cache-Control：用于指定缓存指令，缓存指令是单向的（响应中出现的缓存指令在请求中未必会出现），且是独立的（一个消息的缓存指令不会影响另一个消息处理的缓存机制）。

2．请求报头

请求报头通知服务器关于客户端请求的信息，典型的请求报头如下。

Host：请求的主机名，允许多个域名同处于一个 IP 地址处，即虚拟主机。

User-Agent：发送请求的浏览器类型、操作系统等信息。

Accept：客户端可识别的内容类型列表，用于指定客户端接收哪些类型的信息。

Accept-Encoding：客户端可识别的数据编码。

Accept-Language：表示浏览器所支持的语言类型。

Connection：允许客户端和服务器指定与请求/响应连接有关的选项，如为 Keep-Alive 则表示保持连接。

Transfer-Encoding：告知接收端对报文采用了什么编码方式，用来保证报文的可靠传输。

3．响应报头

响应报头用于服务器传递自身信息的响应，常见的响应报头如下。

Location：用于重定向接收者到一个新的位置，常用在更换域名时。
Server：包含以此服务器处理请求的系统信息，与 User-Agent 请求报头是相对应的。

4．实体报头

实体报头用来定义被传送资源的信息，既可用于请求也可用于响应。请求和响应消息都可以传送一个实体。常见的实体报头如下。

Content-Type：发送给接收者的实体正文的媒体类型。

Content-Lenght：实体正文的长度。

Content-Language：描述资源所用的自然语言，没有设置则认为实体内容将提供给所有的语言阅读。

Content-Encoding：被用作媒体类型的修饰符，它的值指示了已经被应用到实体正文的附加内容的编码，因而要获得 Content-Type 报头域中所引用的媒体类型，必须采用相应的解码机制。

Last-Modified：用于指示资源的最后修改日期和时间。

Expires：给出响应过期的日期和时间。

9.2　Android 网络编程基础

目前 Android 平台有三种网络接口可以使用，分别是 java.net.*（标准 Java 接口）、org.apache（Apache 接口）和 android.net.*（Android 网络接口）。[40]

标准 Java 接口：提供与联网有关的类，包括流和数据包套接字、Internet 协议、常见 HTTP 处理。例如，创建 URL 及 URLConnection/HttpURLConnection 对象、设置连接参数、连接到服务器、向服务器写数据、从服务器读取数据等。[40]

使用 java.net.* 包连接网络：

```
try{
    //定义地址
    URL  url=new URL("http://www.Google.com");
    //打开连接
    HttpURLConnection  http=(httpURLConnection) url.openConnection();
    //得到连接状态
    Inc nrc =http.getResponseCode();
    if (nrc == HttpURLConnection.HTTP_OK) {
        //取得数据
        InputStream is = http.getInputstream();
        //处理数据
    }
}
```

Apache 接口：虽然 JDK 的 java.net 包中已经提供了访问 HTTP 的基本功能，但对于大部分应用程序来说，JDK 库本身提供的功能远远不够，而 Android 提供的 Apache HttpClient 为客户端的 HTTP 编程提供了高效、最新、功能丰富的工具包支持。可以将 Apache 接口视为目前流行的开源 Web 服务器，主要包括创建 HttpClient 及 GET/Post HttpRequest 等对象，设置连接参数，执行 HTTP 操作，处理服务器返回结果等功能。[40]

使用 android.net.http.*连接网络：

```
try
    //创建 HttpClient
    //这里使用 DefaultHttpClient 表示默认属性
    HttpClient  hc=new DefaultHttpClient();
    //HttpGet 实例
    HttpGet get =new HttpGet("http://www.Google.com");
    //连接
    HttpResponse rp =hc.execute(get);
    if(rp.getStatusLine().getStatusCode()==HttpStatus.SC_OK){
        InputStream is =rp.getEntity().getContent();
        //处理数据
    }
}
```

Android 网络接口：Android.net*包实际上是通过对 Apache 中的 HttpClient 的封装来实现一个 HTTP 编程接口的，同时还提供了 HTTP 请求队列管理及 HTTP 连接池管理，以提高并发请求情况下的处理效率。除此之外，还有网络状态监视等接口、网络访问的 Socket、常用的 URL 类及与 WiFi 相关的类等。[40]

Android 中的 Socket 连接：

```
try
//IP 地址
    InetAddress  inetAddress=InetAddress.getByName("192.168.1.110")
    //端口
    Socket client =new socket(inetAddress,61203,ture);
    //取得数据
    InputStream in =client.getInputStream();
    Outstram Out =client.getOutputStream();
    //处理数据
}
```

9.3 HTTP 通信

HTTP（Hyper Transer Protocol，超文本传输协议）用于传送 WWW 形式的数据。HTTP

采用了请求/响应的模型。请求报头包含请求的方法、URL、协议版本，以及请求修饰符、客户信息和内容，类似于 MIME 的消息结构。服务器以一个状态行响应，响应的内容包括消息协议的版本、成功或错误编码，还包括服务器信息、实体元信息及可能的实体内容。Android 提供了 HttpURLConnection 和 HttpClient 接口来开发 HTTP 程序。[41]

9.3.1 HttpURLConnection 接口

GET 请求可以获取静态页面，也可以把参数放在 URL 字符串后面，传递给服务器。POST 与 GET 不同，POST 的参数不是放在 URL 的字符串里，而是放在 HTTP 请求数据中。HttpURLConnection 是 Java 的标准类，继承自 URLConnection。HttpURLConnection 和 URLConnection 都是抽象类，无法直接实例化对象。[41]其对象主要通过 URL 的 openConnection 方法获得。创建一个 HttpURLConnection 连接的代码如下：

```
URL url =new URL ("http://Google.com/");
HttpURLConnection urlConn=(httpURLConnection)url.openConnection();
```

openConnection 方法只创建 URLConnection 或 httpURLConnection 实例，而并不对该实例进行真正的连接操作，并且每次运行 openConnection 都创建一个新的实例。在连接之前，可以对它的一些属性进行设置，如下所示。

```
//设置输入/输出流
connection.setDoOutput(true);
connection.setDoIntput(true);
//设置方式为 POST
connection.setRequestMethod("POST");
//POST 请求不能使用缓存
connection.setUseCathes(false);
//连接完成之后可以关闭这个连接
//关闭 HttpURLConnection 连接
urlConn.disconnect();
```

访问不需要参数的网页的实现如下所示：

```java
public class Activity02 extends Activity{
    private final String DEBUG_TAG = "Activity02";
    /** Called when the activity is first created. */
    @Override
    public void onCreate(Bundle savedInstanceState){
        super.onCreate(savedInstanceState);
        setContentView(R.layout.http);
        TextView mTextView = (TextView)this.findViewById(R.id.Text
        View_HTTP);
        //http 地址
        String httpUrl = "http://192.168.1.110:8080/http1.jsp";
        //获得的数据
        String resultData = "";
```

```java
URL url = null;
try{
//构造一个URL对象
    url = new URL(httpUrl);
}catch (MalformedURLException e){
    Log.e(DEBUG_TAG, "MalformedURLException");
}
if (url != null){
    try{
        //使用HttpURLConnection打开连接
        HttpURLConnection urlConn = (HttpURLConnection)
        url.openConnection();
        //得到读取的内容(流)
        InputStreamReader in = new InputStreamReader(urlConn.
        getInputStream());
        //为输出创建BufferedReader
        BufferedReader buffer = new BufferedReader(in);
        String inputLine = null;
        //使用循环来读取获得的数据
        while (((inputLine = buffer.readLine()) != null)){
            //在每一行后面加上一个"\n"来换行
            resultData += inputLine + "\n";
        }
        //关闭InputStreamReader
        in.close();
        //关闭http连接
        urlConn.disconnect();
        //设置显示取得的内容
        if ( resultData != null ){
            mTextView.setText(resultData);
        }else {
            mTextView.setText("读取的内容为NULL");
        }
    } catch (IOException e){
        Log.e(DEBUG_TAG, "IOException");
    }
}else{
    Log.e(DEBUG_TAG, "Url NULL");
}
//设置按键事件监听
Button button_Back = (Button) findViewById(R.id.Button_Back);
/* 监听button的事件信息 */
button_Back.setOnClickListener(new Button.OnClickListener() {
    public void onClick(View v){
        /* 新建一个Intent对象 */
```

```java
                    Intent intent = new Intent();
                    /* 指定 Intent 要启动的类 */
                    intent.setClass(Activity02.this, Activity01.class);
                    /* 启动一个新的 Activity */
                    startActivity(intent);
                    /* 关闭当前的 Activity */
                    Activity02.this.finish();
                }
            });
        }
    }
```

通过 GET 方式传递参数如下所示：

```java
    //以 GET 方式上传参数
    public class Activity03 extends Activity{
        private final String DEBUG_TAG = "Activity03";
        /** Called when the activity is first created. */
        @Override
        public void onCreate(Bundle savedInstanceState){
            super.onCreate(savedInstanceState);
            setContentView(R.layout.http);
            TextView mTextView = (TextView)this.findViewById(R.id.TextView_HTTP);
            //http 地址"?par=abcdefg"是上传的参数
            String httpUrl = "http://192.168.1.110:8080/httpget.jsp?par=abcdefg";
            //获得的数据
            String resultData = "";
            URL url = null;
            try{
                //构造一个 URL 对象
                url = new URL(httpUrl);
            }catch (MalformedURLException e){
                Log.e(DEBUG_TAG, "MalformedURLException");
            }
            if (url != null){
                try{
                    // 使用 HttpURLConnection 打开连接
                    HttpURLConnection urlConn = (HttpURLConnection)url.openConnection();
                    //得到读取的内容(流)
                    InputStreamReader in = new InputStreamReader(urlConn.getInputStream());
                    // 为输出创建 BufferedReader
                    BufferedReader buffer = new BufferedReader(in);
                    String inputLine = null;
```

```
                    //使用循环来读取获得的数据
                    while (((inputLine = buffer.readLine()) != null)){
                        //在每一行后面加上一个"\n"来换行
                        resultData += inputLine + "\n";
                    }
                    //关闭 InputStreamReader
                    in.close();
                    //关闭 http 连接
                    urlConn.disconnect();
                    //设置显示取得的内容
                    if ( resultData != null ){
                        mTextView.setText(resultData);
                    }else {
                        mTextView.setText("读取的内容为 NULL");
                    }
                } catch (IOException e){
                    Log.e(DEBUG_TAG, "IOException");
                }
            }else{
                Log.e(DEBUG_TAG, "Url NULL");
            }
Button button_Back = (Button) findViewById(R.id.Button_Back);
        /* 监听 button 的事件信息 */
        button_Back.setOnClickListener(new Button.OnClickListener() {
            public void onClick(View v){
                /* 新建一个 Intent 对象 */
                Intent intent = new Intent();
                /* 指定 Intent 要启动的类 */
                intent.setClass(Activity03.this, Activity01.class);
                /* 启动一个新的 Activity */
                startActivity(intent);
                /* 关闭当前的 Activity */
                Activity03.this.finish();
            }
        });
    }
}
```

修改网页地址，加上要传递的参数，如下所示：

```
String httpurl="http://192.168.1.110:8080/httpget.jsp?par=abcdefg";
URL url =new URL(httpUrl);
HttpURLConnection urlConn=(HttpURLConnection)url.openConnection();
```

由于 HTTPURLConnection 会默认使用 GET 方法，所以如果想要使用 POST 方法，则需要进行 setRequestMethod 设置。然后将需要传递的参数通过 writeBytes 方法写入数据流。代码如下所示：

```java
//以POST方式上传参数
public class Activity04  extends Activity{
    private final String DEBUG_TAG = "Activity04";
    /** Called when the activity is first created. */
    @Override
    public void onCreate(Bundle savedInstanceState){
        super.onCreate(savedInstanceState);
        setContentView(R.layout.http);
        TextView mTextView = (TextView)this.findViewById
        (R.id.TextView_HTTP);
        //http地址"?par=abcdefg"是上传的参数
        String httpUrl = "http://192.168.1.110:8080/httpget.jsp";
        //获得的数据
        String resultData = "";
        URL url = null;
        try{
            //构造一个URL对象
            url = new URL(httpUrl);
        }catch (MalformedURLException e){
            Log.e(DEBUG_TAG, "MalformedURLException");
        }
        if (url != null){
            try{
                //使用HttpURLConnection打开连接
                HttpURLConnection urlConn = (HttpURLConnection)
                url.openConnection();
                //因为这个是POST请求,所以需要设置为true
                urlConn.setDoOutput(true);
                urlConn.setDoInput(true);
                //设置方式为POST
                urlConn.setRequestMethod("POST");
                //POST请求不能使用缓存
                urlConn.setUseCaches(false);
                urlConn.setInstanceFollowRedirects(true);
                // 配置本次连接的 Content-type, 配置为
                //application/x-www-form-urlencoded的连接,从postUrl.
                //openConnection()至此的配置必须要在connect之前完成,
                //要注意的是connection.getOutputStream会隐含地进行
                //connect
                urlConn.setRequestProperty("Content-Type","
                application/x-www-form-urlencoded");
                urlConn.connect();
                //DataOutputStream流
                DataOutputStream out = new DataOutputStream
```

```
                    (urlConn.getOutputStream());
                //要上传的参数
                String content="par="+URLEncoder.encode
                ("ABCDEFG", "gb2312");
                //将要上传的内容写入流中
                out.writeBytes(content);
                //刷新、关闭
                out.flush();
                out.close();
                //获取数据
                BufferedReader reader = new BufferedReader(new
                InputStreamReader(urlConn.getInputStream()));
                String inputLine = null;
                //使用循环来读取获得的数据
                while (((inputLine = reader.readLine()) != null)){
                    //在每一行后面加上一个"\n"来换行
                    resultData += inputLine + "\n";
                }
                reader.close();
                //关闭http连接
                urlConn.disconnect();
                //设置显示取得的内容
                if ( resultData != null ){
                    mTextView.setText(resultData);
                }else {
                    mTextView.setText("读取的内容为NULL");
                }
            }catch (IOException e){
                Log.e(DEBUG_TAG, "IOException");
            }
        }else{
            Log.e(DEBUG_TAG, "Url NULL");
        }
        Button button_Back = (Button) findViewById(R.id.Button_Back);
        /* 监听button的事件信息 */
        button_Back.setOnClickListener(new Button.OnClickListener() {
            public void onClick(View v){
                Intent intent = new Intent();
                intent.setClass(Activity04.this, Activity01.class);
                startActivity(intent);
                Activity04.this.finish();
            }
        });
    }
}
```

如果在开发中想从网络上获取一张图片进行显示，连接方式与前述内容相同，只是需要将连接之后得到的数据流转换成 Bitmap 形式。取得网络上图片的代码如下所示：

```
//url :图片地址
Public Bitmap GetNetBitmap(String url){
    URL imageUrl=null;
    Bitmap bitmap=null;
    try{
        Imageurl=new URL(url);
    }Catch(MalformedURLException e){
        Log.e(DEBUG_TAG,e.getMessage());
    }
    try{
        HTTPURLConnection conn=(HttpURLConnection)
         imageUrl.openConnection();
        conn.setDoInput(true);
        Conn.connect();
        //将得到的数据转换成 Inputstream
        Inputstream is = conn.getInputStream();
        //将 jiangInputstream 转换为 Bitmap
        Bitmap =BitmapFactory.decodeStream(is);
        Is.close;
    }Catch(IOException e){
        Log.e(DEBUG_TAG,e.getMessage());
    }
    Return bitmap;
}
```

9.3.2 HttpClient 接口

HttpClient 是 Apache 对 Java 中的 HttpURLClient 接口的封装，主要引用 org.apache.http.**。Google 在 2.3 版本之前推荐使用 HttpClient，因为这个封装包的安全性高。ClientConnectionManager 接口是客户端连接管理器接口，提供的抽象方法如表 9-3 所示。

表 9-3 ClientConnectionManager 接口的抽象方法

方法	说明
ClientConnectionManager	关闭所有无效、超时的连接
closeIdleConnections	关闭空闲的连接
releaseConnection	释放一个连接
RequestConnection	请求一个新的连接
shutdown	关闭管理器并释放资源

DefaultHttpClient 是一个默认的 HTTP 客户端，能够使用它建立一个 HTTP 连接：

```
httpClient httpclient =new DefaultHttpClient();
```

HttpResponse 是 HTTP 连接后的响应，当程序执行一个 HTTP 连接后，就会返回一个 HttpResponse，可以通过分析 HttpResponse 来获取一些响应的信息。

下面是请求 HTTP 连接并且获取该请求是否成功的代码：

```
HttpResponse httpResponse =httpclient.execute(httpRequest);
if(httpResponse.getStatusLine().getStatusCode()==HttpStatus.SC_OK){
}
```

HttpClient 使用 GET 方式取得数据：

```
public class Activity02 extends Activity{
    /** Called when the activity is first created. */
    @Override
    public void onCreate(Bundle savedInstanceState){
        super.onCreate(savedInstanceState);
        setContentView(R.layout.http);
        TextView mTextView = (TextView) this.findViewById
        (R.id.TextView_HTTP);
        //http 地址
        String httpUrl = "http://192.168.1.110:8080/httpget.jsp?par
        =HttpClient_ android_Get";
        //HttpGet 连接对象
        HttpGet httpRequest = new HttpGet(httpUrl);
        try{
            //取得 HttpClient 对象
            HttpClient httpclient = new DefaultHttpClient();
            //请求 HttpClient，取得 HttpResponse
            HttpResponse httpResponse = httpclient.execute(httpRequest);
            //请求成功
            if (httpResponse.getStatusLine().getStatusCode() == Http
            Status.SC_OK){
            //取得返回的字符串
                String strResult = EntityUtils.toString(httpResponse.
                getEntity());
                mTextView.setText(strResult);
            }else{
                mTextView.setText("请求错误!");
            }
        } catch (ClientProtocolException e){
            mTextView.setText(e.getMessage().toString());
        }catch (IOException e){
            mTextView.setText(e.getMessage().toString());
        }catch (Exception e){
            mTextView.setText(e.getMessage().toString());
        }
        //设置按键事件监听
```

```java
            Button button_Back = (Button) findViewById(R.id.Button_Back);
            /* 监听 button 的事件信息 */
            button_Back.setOnClickListener(new Button.OnClickListener() {
                public void onClick(View v){
                    /* 新建一个 Intent 对象 */
                    Intent intent = new Intent();
                    /* 指定 Intent 要启动的类 */
                    intent.setClass(Activity02.this, Activity01.class);
                    /* 启动一个新的 Activity */
                    startActivity(intent);
                    /* 关闭当前的 Activity */
                    Activity02.this.finish();
                }
            });
        }
    }
```

使用 POST 方式取得数据：

```java
public class Activity03 extends Activity{
    /** Called when the activity is first created. */
    @Override
    public void onCreate(Bundle savedInstanceState){
        super.onCreate(savedInstanceState);
        setContentView(R.layout.http);
        TextView mTextView = (TextView) this.findViewById
        (R.id.TextView_HTTP);
        //http 地址
        String httpUrl = "http://192.168.1.110:8080/httpget.jsp";
        //HttpPost 连接对象
        HttpPost httpRequest = new HttpPost(httpUrl);
        //使用 NameValuePair 来保存要传递的 POST 参数
        List<NameValuePair> params = new ArrayList<NameValuePair>();
        //添加要传递的参数
        params.add(new BasicNameValuePair("par", "HttpClient_
        android_Post"));
        try{
            //设置字符集
            HttpEntity httpentity=new UrlEncodedFormEntity(params,
            "gb2312");
            //请求 httpRequest
            httpRequest.setEntity(httpentity);
            //取得默认的 HttpClient
            HttpClient httpclient = new DefaultHttpClient();
            //取得 HttpResponse
            HttpResponse httpResponse = httpclient.execute(httpRequest);
```

```
                //HttpStatus.SC_OK 表示连接成功
                if (httpResponse.getStatusLine().getStatusCode() ==
                HttpStatus.SC_OK){
                    //取得返回的字符串
                    String strResult = EntityUtils.toString(httpResponse.
                    getEntity());
                    mTextView.setText(strResult);
                }else{
                    mTextView.setText("请求错误!");
                }
            }catch (ClientProtocolException e){
                mTextView.setText(e.getMessage().toString());
            }catch (IOException e){
                mTextView.setText(e.getMessage().toString());
            }catch (Exception e){
                mTextView.setText(e.getMessage().toString());
            }
            //设置按键事件监听
            Button button_Back = (Button) findViewById(R.id.Button_Back);
            /* 监听 button 的事件信息 */
            button_Back.setOnClickListener(new Button.OnClickListener() {
                public void onClick(View v){
                    /* 新建一个 Intent 对象 */
                    Intent intent = new Intent();
                    /* 指定 Intent 要启动的类 */
                    intent.setClass(Activity03.this, Activity01.class);
                    /* 启动一个新的 Activity */
                    startActivity(intent);
                    /* 关闭当前的 Activity */
                    Activity03.this.finish();
                }
            });
        }
    }
```

9.3.3 实时更新

实时更新是指通过一个线程来控制视图的更新。如果想要实现从网络中实时获取数据，就需要把获取网络数据的代码写入线程中，不停地进行更新。

举例说明：

```
    public class Activity01 extends Activity{
        private final String DEBUG_TAG = "Activity02";
        private TextView mTextView;
```

```java
        private Button mButton;
        /** Called when the activity is first created. */
        @Override
        public void onCreate(Bundle savedInstanceState){
            super.onCreate(savedInstanceState);
            setContentView(R.layout.main);
            mTextView = (TextView)this.findViewById(R.id.TextView01);
            mButton = (Button)this.findViewById(R.id.Button01);
            mButton.setOnClickListener(new Button.OnClickListener(){
                @Override
                public void onClick(View arg0){
                    // TODO Auto-generated method stub
                    refresh();
                }
            });

            //开启线程
            new Thread(mRunnable).start();
        }
        //刷新网页显示
        private void refresh(){
            String httpUrl = "http://192.168.1.110:8080/date.jsp";
            String resultData = "";
            URL url = null;
            try{
                //构造一个URL对象
                url = new URL(httpUrl);
            }catch (MalformedURLException e){
                Log.e(DEBUG_TAG, "MalformedURLException");
            }
            if (url != null){
                try{
                    //使用HttpURLConnection打开连接
                    HttpURLConnection urlConn=(HttpURLConnection)
                    url.openConnection();
                    //得到读取的内容(流)
                    InputStreamReader in = new InputStreamReader
                    (urlConn.getInputStream());
                    //为输出创建BufferedReader
                    BufferedReader buffer = new BufferedReader(in);
                    String inputLine = null;
                    //使用循环来读取获得的数据
                    while (((inputLine = buffer.readLine()) != null)){
```

```java
                //在每一行后面加上一个"\n"来换行
                resultData += inputLine + "\n";
            }
            //关闭InputStreamReader
            in.close();
            //关闭http连接
            urlConn.disconnect();
            //设置显示取得的内容
            if (resultData != null){
                mTextView.setText(resultData);
            }else{
                mTextView.setText("读取的内容为NULL");
            }
        }catch (IOException e){
            Log.e(DEBUG_TAG, "IOException");
        }
    }else{
        Log.e(DEBUG_TAG, "Url NULL");
    }
}
private Runnable mRunnable = new Runnable(){
    public void run(){
        while (true){
            try{
                Thread.sleep(5 * 1000);
                //发送消息
                mHandler.sendMessage(mHandler.obtainMessage());
            } catch (InterruptedException e){
                // TODO Auto-generated catch block
                Log.e(DEBUG_TAG, e.toString());
            }
        }
    }
}
Handler mHandler = new Handler(){
    public void handleMessage(Message msg){
        super.handleMessage(msg);
        //刷新
        refresh();
    }
}
```

9.4 Socket 通信

Socket 翻译过来是"插座"的意思，计算机专业术语则称之为"套接字"，用于描述 IP 地址和端口（可以从 Socket 的初始化语句角度理解），是一个通信链的句柄（句柄是一个标识符，用来标识对象或项目），可以用来实现不同虚拟机或不同计算机之间的通信。应用程序通常通过"套接字"向网络发出请求或应答网络请求。[42]

Socket 类似于电话插座。以一个国家级电话网为例，电话的通话双方相当于相互通信的两个进程，区号是它的网络地址；区内一个单位的交换机相当于一台主机，主机分配给每个用户的局内号码相当于 Socket 号。任何用户在通话之前，首先要占有一部电话机，相当于申请一个 Socket；同时要知道对方的号码，相当于对方有一个固定的 Socket。然后向对方拨号呼叫，相当于发出连接请求（假如对方不在同一区内，还要拨对方的区号，相当于给出网络地址）；对方假如在场并空闲（相当于通信的另一主机开机且可以接受连接请求），拿起电话话筒，双方就可以正式通话（相当于连接成功）双方通话的过程是一方向电话机发出信号和对方从电话机接收信号的过程，相当于向 Socket 发送数据和从 Socket 接收数据。通话结束后，一方挂起电话机相当于关闭 Socket，撤销连接。[42]

Socket 是通信的基石，是支持 TCP/IP 协议的网络通信的基本单元，是网络通信过程中端点的抽象表示，包含网络通信必需的 5 种信息：连接使用的协议、本地主机的 IP 地址、本地进程的协议端口、远地主机的 IP 地址和远地进程的协议端口。[42]

9.4.1 Socket 传输模式

Socket 有两种主要的操作方式：面向连接的和面向非连接的。到底采用哪种模式，要看应用程序的需要，如果需要可靠、安全的传输则选择面向连接的，否则选择无连接。无连接的操作使用数据报协议，这个模式下的 Socket 不需要建立连接，只是简单地投出数据报。它是快速的、高效的，但是数据安全性不佳。面向连接的操作使用 TCP 协议。在这个模式下，Socket 必须在发送数据之前与目的地的 Socket 取得连接，一旦建立连接，Socket 就可以使用一个流接口进行打开、读、写、关闭操作，所有发送信息都会在另一端以同样的顺序被接收。[43]

Socket 编程原理如下。

1. Socket 构造

Java 在 java.net 中提供了两个类，分别是 Socket 类和 ServerSocket 类，分别用来表示双向连接的客户端和服务端。ServerSocket 用于服务器端，Socket 是建立网络连接时使用的。当连接成功后，应用程序两端都会产生一个 Socket 实例，操作这个实例，完成所需的会话。

对于一个网络连接来说，套接字是平等的，并没有差别，不会因为在服务器端或在客户端而产生不同级别。

重要的 Socket API：java.net.Socket 继承于 java.lang.Object，有 8 个构造器，其方法不多。

Accept 方法用于产生"阻塞"，直到接收到一个连接，并且返回一个客户端的 Socket 对象实例。"阻塞"是一个术语，它使程序运行暂时"停留"在这个地方，直到一个会话产生，然后程序继续；通常"阻塞"是由循环产生的。

getInputStream 方法获得网络连接输入，同时返回一个 InputStream 对象实例。

getOutputStream 方法连接的另一端将得到输入，同时返回一个 OutputStream 对象实例。

注意：getInputStream 和 getOutputStream 方法均可能会产生一个 IOException，它必须被捕获，因为它们返回的流对象通常都会被另一个流对象使用。[44]

构造方法如下：

Socket（InetAddress address，int port）；

Socket（InetAddress address，int port，boolean stream）；

Socket（String host，int port）；

Socket（String host，int port，boolean stream）；

Socket（SocketImpl impl）；

Socket（String host，int port，IneAddress localAddr，int localPort）；

Socket（InetAddress address，int port，IneAddress localAddr，int localPort）；

ServerSocket（int port，int backlog）；

ServerSocket（int port，int backlog，InetAddress bindAddr）；

address、host、port 分别是双向连接中另一方的 IP 地址、主机名和端口号。

localPort 表示本地主机的端口号，localAddr 和 bindAddr 是本地机器的地址（ServerSocket 的主机地址），impl 是 Socket 的父类，既可以用来创建 ServerSocket，又可以用来创建 Socket。Count 则表示服务端所能支持的最大连接数。

2．客户端 Socket

在客户端创建一个 Socket，并指出需要连接的服务器的 IP 地址和端口。

3．ServerSocket

创建服务器端 SeverSocket。

4．输入/输出流

Socket 为开发者提供了 getInputStream()和 getOutStream()方法来对对应的输入（出）流进行读写操作，这两个方法分别会返回 InputStream 和 OutputStream 类对象。为了便于对数据进行读写操作，开发者可以在返回的输入/输出流对象上同时建立过滤流，如 DataInputStram、DataOutputStream 或 PrintStream 类对象。对于文本方式流的对象，可以采用 InputStreamReader 和 OutputStreamWriter、PrintWriter 等处理。

5. 关闭 Socket 和流

关闭 Socket 需要调用 Socket 的 close()方法。在关闭之前，应该将与 Socket 有关的所有输入/输出流全部关闭，以释放所有的资源，如下所示：

os. close();
　　is.close();
　　socket.close();

9.4.2　Android Socket 编程步骤

服务器端实现：

（1）指定端口对 ServerSocket 进行实例化；
（2）调用 ServerSocket 的 acceptv()方法以防在连接期间造成阻塞；
（3）获取底层 Socket 流并对其进行读写操作；
（4）将数据封装成流；
（5）对 Socket 进行读写；
（6）关闭所有打开的输入/输出流。

代码的主要内容如下所示：

```java
import java.io.BufferedReader;
import java.io.BufferedWriter;
import java.io.InputStreamReader;
import java.io.OutputStreamWriter;
import java.io.PrintWriter;
import java.net.ServerSocket;
import java.net.Socket;
public class Server implements Runnable{
    public void run(){
        try{
            //创建 ServerSocket 在端口 54321 监听客户请求
            ServerSocket  serverSocket = new ServerSocket(54321);
            while (true){
                //接收客户端请求
                Socket client = serverSocket.accept();
                System.out.println("accept");
                try{
                    //接收客户端消息
                    BufferedReader in = new BufferedReader(new
                    InputStreamReader(client.getInputStream()));
                    String str = in.readLine();
                    System.out.println("read:" + str);
                    //向服务器发送消息
                    PrintWriter out = new PrintWriter( new BufferedWriter
```

```java
            ( new OutputStreamWriter(client.getOutputStream())),
                true);
                out.println("server message");
                //关闭流
                out.close();
                in.close();
            }catch (Exception e){
                System.out.println(e.getMessage());
                e.printStackTrace();
            }finally{
                //关闭
                client.close();
                System.out.println("close");
            }
        }
        }catch (Exception e){
            System.out.println(e.getMessage());
        }
    }
    //main函数,开启服务器
    public static void main(String a[]){
        Thread desktopServerThread = new Thread(new Server());
        desktopServerThread.start();
    }
}
```

客户端实现:

(1) 通过 IP 地址和端口实例化 Socket,请求连接服务器;

(2) 获取 Socket 上的流以进行读写;

(3) 把流包进 BufferedReader /PrintWriter 的实例;

(4) 对 Socket 进行读写;

(5) 关闭打开的流。

代码的主要内容如下所示:

```java
public class Activity01 extends Activity{
    private final String DEBUG_TAG = "Activity01";
    private TextView mTextView = null;
    private EditText mEditText = null;
    private Button  mButton = null;
    /** Called when the activity is first created. */
    @Override
    public void onCreate(Bundle savedInstanceState){
        super.onCreate(savedInstanceState);
        setContentView(R.layout.main);
        mButton=(Button)findViewById(R.id.Button01);
```

```java
mTextView=(TextView)findViewById(R.id.TextView01);
mEditText=(EditText)findViewById(R.id.EditText01);
//登录
mButton.setOnClickListener(new OnClickListener(){
    public void onClick(View v){
        Socket socket = null;
        String message = mEditText.getText().toString() + "\r\n";
        try {
            //创建 Socket
            socket = new Socket("192.168.1.110",54321);
            //向服务器发送消息
            PrintWriter out=new PrintWriter(newBufferedWriter(newOutputStreamWriter
            (socket.getOutputStream())),true);
            out.println(message);
            //接收来自服务器的消息
            BufferedReader br=new BufferedReader(new InputStreamReader(socket.getInputStream()));
            String msg = br.readLine();
            if ( msg != null ){
                mTextView.setText(msg);
            }else{
                mTextView.setText("数据错误!");
            }
            //关闭流
            out.close();
            br.close();
            //关闭 Socket
            socket.close();
        }catch (Exception e) {
            // TODO: handle exception
            Log.e(DEBUG_TAG, e.toString());
        }
    }
}};
    }
}
```

9.5 Socket 应用

简易的聊天室是 Socket 应用之一。在实际应用过程中，一般是在服务器上运行一个永久性的程序，它可以接收来自其他多个客户端的请求，并提供相应的服务。为了实现应用，需利用多线程来实现多客户机制。

服务器在指定的端口上监听是否有客户请求，一旦程序监听到有客户请求，服务器就会启动一个新线程来响应该客户的请求，而服务器本身在启动完此线程后就马上进入监听状态，等待下一个客户请求的到来。

9.6 WebKit 应用

WebKit 是一个开源浏览器网页排版引擎。它由 3 个模块组成：JavaScriptCore（JavaScript 解释器）、WebCore（整个项目的核心，用来实现 Render 引擎，解析 Web 页面，生成一个 dom 树和一个 Render 树）、Webkit（整个项目的名称）。

WebKit 工作流程如下。

（1）用户向 Shell 发出页面请求后，页面的 URL 或本地文件名被发送到 Shell。

（2）Shell 调用 IO 组件，把 URL 传送到 IO 组件。

（3）IO 组件使用 HTTP 或调用本地 IO 获取 HTML/XHTML 源数据，返回 Shell。

（4）Shell 把 IO 返回的 HTML/XHTML source 提交给 HTML/XHTML 分析器。

（5）HTML/XHTML 分析器分析 HTML/XHTML 代码，构建一棵 DOM 树，树根为 HTML-Document。

（6）通过 DOM 树，生成 Render 树。简单来说，它是对 DOM 树更进一步的描述，其描述的内容主要与布局渲染等 CSS 相关属性（如 left、top、width、height、cofont 等）有关，因为不同的 DOM 树节点可能会有不同的布局渲染属性，甚至布局时会按照标准动态生成一些匿名节点。为了更加方便地描述布局及渲染，WebKit 内核又生成一棵 Render 树来描述 DOM 树的布局渲染等特性。当然，DOM 树与 Render 树不是一一对应的，但可以相互关联。

（7）布局器布局显示出来。当布局管理器对可视化元素指派好位置、大小后，它就知道了它的安身之处，也记住了它的大小，它必须严格遵守布局管理器给它分配的位置、大小，不能擅自更改。既然知道了自己的位置、大小，剩下的就是控件根据自己的属性表现自己的背景、外形等。[45]

流程图如图 9-2 所示。

WebKit 的解析过程：

（1）CURL 获得网站的 stream；

（2）解析划分字符串；

（3）通过 DOM Builder 按合法的 HTML 规范生成 DOM 树；

（4）把 DOM 传给 layoutEngine 进行布局，如果有 CSS 样式，就通过 CSSParser 解析；

（5）最后通过 Rendering 渲染出来。

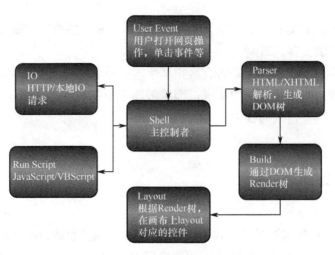

图 9-2 WebKit 流程图

Google 对 WebKit 进行了封装，为开发者提供了丰富的 Java 接口，其中最重要的是 android.Webkit WebView 控件。WebView 控件专门用来浏览网页。需要在 XML 布局文件中定义一个 WebView 控件，然后在程序中装载它，设置其属性，如颜色、字体、要访问的网址等。Android 中专门通过 WebSeTings 来设置 WebView 的一些属性状态。创建 WebView 控件时，系统会设置一个默认值，开发者可以通过 WebView.getSeTings 方法来得到这个设置值。[45]

9.7 WiFi 编程

WiFi 的全称是 Wireless Fidelity，是无线局域网联盟（WLANA）的商标，该商标仅保障使用该商标的商品互相之间可以合作。WiFi 最开始是遵从 IEEE 802.11b 标准的一种通信技术，和蓝牙同级，因此后来人们习惯用 WiFi 来称呼 802.11b 协议。它的最大优点就是传输速度高，可以达到 11Mbps，并且它的有效距离也很长，可达几十米，而且它与已有的各种 802.11 DSSS 设备兼容。因此，WiFi 一直是企业实现内部无线局域网的主要技术。

一些常用的操作如下所示。

ScanResult：主要用来描述已经检测出来的接入点。

WifiConfiguration：WiFi 网络的配置。

WifiInfo：WiFi 无线连接的描述。

Wifi.Manager：提供了管理 WiFi 的大部分 API。

WifiManager.Wifilock：允许应用程序一直使用 WiFi 无线网络。通常情况下，当用户在一段时间内没有进行任何操作时，WiFi 网络就会自动关闭，但是通过 Wifilock 锁定后，该段时间就会一直保持连接。

WifiManager 的常用方法如表 9-4 所示。

表 9-4 WifiManager 方法

方 法	说 明
addNetwork	添加一个配置好的网络连接
calculateSignalLevel	计算信号的强度
ComparesignalLevel	比较两个信号的强度
creatWifiLock	创建一个 WiFi 锁
disableNetwork	取消一个配置好的网络,使其不可用
disconnect	从接入点断开
enableNetwork	允许指定的网络连接
getConfigureNetworks	得到客户端所有已经配置好的网络列表
getConnectionInfo	得到正在使用的连接的动态信息
getDhcpInfo	得到最后一次成功的 DHCP 请求的 DHCP 地址
getScanResult	得到被扫描的接入点
getwifistate	得到可用的 WiFi 的状态
iswifiEnabled	检查 WiFi 是否可用
pingSupplicant	检查客户端对请求的响应
ressociate	从当前接入点重新连接
removeNetwork	从已经配置好的网络列表中删除指定 ID 的网络
saveconfiguration	保存当前配置好的网络列表
setwifiEnabled	设置 WiFi 是否可用
starscan	扫描存在的接入点
updataNetwork	更新已经配置好的网络

9.8 蓝牙编程

蓝牙的类和接口位于 android.bluetooth 包中,蓝牙包提供的功能如表 9-5 所示。

表 9-5 蓝牙包功能

功 能 包	说 明
BluetoothAdapter	蓝牙适配器
BluetoothClass	蓝牙类(主要包括服务和设备)
BluetoothClass.Device	蓝牙设备类
BluetoothClass.Device.Major	蓝牙设备管理
BluetoothClass.Service	有关蓝牙服务的类
BluetoothDevice	蓝牙设备(远程蓝牙设备)
BluetoothServerSocket	监听蓝牙连接的类
BluetoothSocket	蓝牙连接类

bluetoothAdapter 中的动作常量如表 9-6 所示。

表 9-6 bluetoothAdapter 中的动作常量

动作常量	说明
ACTION_DISCOVERY_FINISHED	已完成蓝牙搜索
ACTION_DISCOVERY_STARTED	已经开始搜索蓝牙设备
ACTION_LOCAL_NAME_CHANGED	更改蓝牙的名字
ACTION_REQUEST_DISCOVERYABLE	请求能够被搜索
ACTION_REQUEST_ENABLE	请求启动蓝牙
ACTION_SCAN_MODE_CHANGED	扫描模式已改变
ACTION_STATE_CHANGED	状态已改变

bluetoothAdapter 中的常用方法如表 9-7 所示。

表 9-7 bluetoothAdapter 中的常用方法

方法	说明
canceDiscovery	取消当前设备搜索的过程
checkBluetoothAddress	检查蓝牙地址是否正确，如'00：43：A8:23:10:F0'
disable	关闭蓝牙适配器
enable	打开蓝牙适配器
getAddress	取得本地蓝牙的硬件适配器地址
getDefaultAdapter	得到默认的蓝牙适配器
getName	得到蓝牙的名字
getRemoteDevice	得到指定蓝牙硬件地址的 BluetoothDevice 对象
getScanMode	得到扫描模式
getState	得到状态
isDiscovering	是否允许被搜索
isEnabled	是否打开
setName	设置名字
startDiscovery	开始搜索

蓝牙 API 允许应用程序连接和断开蓝牙耳机、扫描仪和其他蓝牙设备，对其进行编写和修改本地服务的 SDP 协议数据库、查询其他蓝牙设备上的 SDP 协议数据库等操作，在 Android 上建立 RFCOMM 协议的连接并连接到其指定的设备上。

第10章 AndroidOpenGL 应用开发

10.1 AndroidOpenGL ES

OpenGL 的英文全称为 OpenGraphicsLibrary，它定义了一个跨编程语言、跨平台的编程接口规格，是一个性能卓越的三维图形标准，并且提供了一个专业的图形程序接口，具有功能强大、调用方便的底层图形库。虽然目前 DirectX 较为常用，但在专业高端绘图领域，OpenGL 是最好的选择，它现在主要用于实现 3D 图形绘制。[46]

OpenGL 是计算机上的大型绘图应用，而在手持移动终端的绘图领域，由于硬件的局限，在绘图时使用简化版的 OpenGL，那就是 OpenGL ES。OpenGL ES 是专为内嵌和移动终端设备设计的 2D /3D 轻量图形库，是基于 OpenGL 的三维图形 API 的子集，目前主要应用于手机、PDA、游戏机等嵌入式设备中。OpenGL ES 的版本有两个：OpenGL ES 1.x（针对固定管线硬件）和 OpenGL ES 2.x（针对可编程管线硬件）。

Android SDK 系统提供了各种绘图的类，除了前面介绍的基本的 Graphics 2D 外，还有更为绚丽的 3D 效果，在很多现有的 Android 系统手机的操作界面中都有所体现，这也是 Android 系统吸引用户的地方。[46]

10.1.1 构建 OpenGL 基本框架

第一步，在 Android 系统中创建一个 3D 开发基本构架，直接导入下面的 Java 库：
```
import javax.microedition.khronos.opengles.GL10;
```
第二步，通过创建 OpenGLContext 对象，直接实例化 GL10（10 表示 1.0 版本），而直接实例化 OpenGLContext 对象 myOpenGLContext 就创建了一个支持 3D 的接口。例如：
```
GL10 gl = (GL10)(myOpenGLContext.getGL());
```
第三步，在重定义 onDraw 方法中处理 Canvas 时需要分别调用 Canvas 绘制的 myOpenGL.Context.waitNative 和 waitG()作为开始和结束。

10.1.2 OpenGL 视图显示

Android 系统提供了 GLSurfaceView 来显示 OpenGL 视图，其中包含一个专门用来渲染 3D 效果的 Renderer 接口。开发者可以自定义一个 Renderer 类，只是需要引入下面的接口：

```
import android.opengl.GLSurfaceView.Renderer;
```

然后再创建一个类。

```
myGLRender implements Renderer
{   ...
    //下面是必须实现的三个抽象方法
    public void onDrawFrame(GL10 gl){}
    public void onSurfaceCreated(GL10 gl, EGLConfig config){}
    public void onSurfaceChanged(GL10 gl, int width, int height){}
    ...
}
```

OpenGL 视图绘制步骤如下。

（1）onDrawFrame 主要应用在 OpenGL 的绘制上，可以绘制各种具体的图形图像，绘制前需要清除屏幕和进行深度缓存。

```
gl.glClear(GL10.GL_COLOR_BUFFER_BIT | GL10.GL_DEPTH_BUFFER_BIT);
gl.glLoadIdentity();// 重置当前的模型观察矩阵，坐标原点为屏幕中心
//绘图例子:设置和绘制正方形
gl.glVertexPointer(3, GL10.GL_FIXED, 0, quaterBuffer);
gl.glDrawArrays(GL10.GL_TRIANGLE_STRIP, 0, 4);
```

（2）onSurfaceCreated 用于对 OpenGL 进行初始化设置。

函数原型：

```
glShadeModel(intmode)
glClearColor(floatred, floatgreen, floatblue, floatalpha)
glClearDepthf(floatdepth)
glEnable(intcap)
glEnable(int cap)
glDepthFunc(int func)
glHint(int target, int mode)
```

代码如下所示：

```
gl.glShadeModel(GL10.GL_SMOOTH);              // 启用阴影平滑
gl.glClearColor(0, 0, 0, 0);                  // 黑色背景
gl.glClearDepthf(1.0f);                       // 设置深度缓存
gl.glEnable(GL10.GL_DEPTH_TEST);              // 启用深度测试
gl.glDepthFunc(GL10.GL_LEQUAL);               // 所作深度测试的类型
gl.glHint(GL10.GL_PERSPECTIVE_CORRECTION_HINT,GL10.GL_FASTEST);
```

（3）onSurfaceChanged 的详细设置。

设置视窗大小的代码如下所示：

```
glViewport (int x, int y, int width, int height)
```
设置投影矩阵和模型观察矩阵的代码如下所示：
```
glMatrixMode (int mode)          //设置投影矩阵
glFrustumf (float left, float right, float bottom, float top, float zNear, float zFar)
```
例如：
```
gl.glMatrixMode(GL10.GL_PROJECTION);
gl.glLoadIdentity();       //重置投影矩阵,调用后三位坐标原点在屏幕中心
gl.glFrustumf(-ratio, ratio, -1, 1, 1, 10);          // 设置视窗的大小
gl.glMatrixMode(GL10.GL_MODELVIEW);                  // 选择模型观察矩阵
gl.glLoadIdentity();       // 重置模型观察矩阵
```

（4）应用自定义的类。定义好自己的 myGLRender 类后，要进行的是在 Activity 中用 setRender 方法把自定义的 myGLRender 类设置为默认的 Renderer，然后通过 setContentView 启动 GLSurfaceView 视图。

代码如下所示：
```
Renderer render = new myGLRender();          //声明 myGLRender 对象
GLSurfaceView glView = new GLSurfaceView(this);
glView.setRenderer(render);          //设置为默认的 Renderer
setContentView(glView);              //启动显示 OpenGL 视图
```

10.2 OpenGL 的三维坐标基础

前面使用过的所有屏幕坐标都是二维坐标，原点位于屏幕左上角。在 Android 系统中，当 OpenGL 对象调用 glLoadIdentity()之后就将当前点移到屏幕中心，没有参数，当需要将中心移到另一个位置时，用 OpenGL 对象调用 glTranslatef（X，Y，Z）可实现偏移屏幕中心具体的位置，其中 X、Y、Z 表示相对屏幕中心的偏移量。

当 OpenGL 对象调用 glLoadIdentity 内置函数后，屏幕中心为坐标原点（X 坐标轴为从左至右，Y 坐标轴为从下至上，Z 坐标轴为从里向外）。每一个坐标点都有三个对应的坐标值（X，Y，Z），再回三维图形时只需要在坐标值上进行变化移动即可。

10.3 多边形的绘制及其颜色渲染

学习了搭建 OpenGl 的基本框架，现在来看看具体的绘图（简单多边形）及渲染例子。绘图和渲染文件如下所示（完整源代码：AndroidDevelopment\Chapter10\ Section10_1 \OpenGL_Polygon_Color\OpenGL_polygon_color）：
```
package com.OpenGL_polygon_color;
```

```java
import java.nio.IntBuffer;            //Java 数组缓冲区
import javax.microedition.khronos.egl.EGLConfig;
import javax.microedition.khronos.opengles.GL10;
import android.opengl.GLSurfaceView.Renderer;
public class MyGLRender implements Renderer{
    int one = 0x10000;//定义整型常量
    float rotateTri,rotateQuad;
    //三角形的3个顶点
    private IntBuffer triggerBuffer = BufferUtil.fBuffer(new int[]{
        0,3*one,0,          //上顶点
        -2*one,one,0,       //左下点
        one,one,0,});       //右下点
    //不规则四边形的4个顶点
    private IntBuffer quaterBuffer = BufferUtil.fBuffer(new int[]{
        2*one,4*one,0,
        -one,4*one,0,
        one,one,0,
        -one,one,0,});
    //#########下面是具有三维坐标顶点颜色渲染的三角形的3个顶点##############
    private IntBuffer triggerBuffer1 = BufferUtil.fBuffer(new int[]{
        -one,-one,0,            //上顶点
        -2*one,-4*one,-one,     //左下点
        one,-3*one,0,});        //右下点
    //有三维形式矩形的4个顶点
    private IntBuffer quaterBuffer1 = BufferUtil.fBuffer(new int[]{
        2*one,-one,0,
        -one,-one,0,
        2*one,-3*one,0,
        -one,-3*one,one});
    //三角形的顶点颜色值(red,green,blue,alpha)
    private IntBuffer colorBuffer = BufferUtil.fBuffer(new int[]{
        one,0,0,one,
        0,one,0,one,
        0,0,one,one,});
    @Override
    public void onDrawFrame(GL10 gl){
        //清除屏幕和深度缓存
        gl.glClear(GL10.GL_COLOR_BUFFER_BIT |
        GL10.GL_DEPTH_BUFFER_BIT);
        //重置当前的模型观察矩阵
```

```
gl.glLoadIdentity();
//左移 1.5 单位,并移入屏幕 6.0
gl.glTranslatef(-1.5f, 0.0f, -6.0f);
gl.glRotatef(rotateTri, 0.0f, 1.0f, 0.0f);
//设置定点数组
gl.glEnableClientState(GL10.GL_VERTEX_ARRAY);
//设置三角形
gl.glVertexPointer(3, GL10.GL_FIXED, 0, triggerBuffer);
//绘制三角形
gl.glDrawArrays(GL10.GL_TRIANGLES, 0, 3);
//重置当前的坐标原点到屏幕中心
gl.glLoadIdentity();
//左移 2.0 单位,并移入屏幕 8.0
gl.glTranslatef(2.0f, 0.0f, -8.0f);
gl.glRotatef(rotateTri, 0.0f, 1.0f, 0.0f);
//设置和绘制不规则四边形
gl.glVertexPointer(3, GL10.GL_FIXED, 0, quaterBuffer);
gl.glDrawArrays(GL10.GL_TRIANGLE_STRIP, 0, 4);
gl.glRotatef(rotateQuad, 1.0f, 0.0f, 0.0f);
//#########三维坐标及颜色渲染###
//重置当前的坐标原点到屏幕中心
gl.glLoadIdentity();
//左移 1.5 单位,并移入屏幕 6.0
gl.glTranslatef(-2.0f, 0.0f, -6.0f);
gl.glRotatef(rotateTri, 0.0f, 1.0f, 0.0f);
//设置定点数组
gl.glEnableClientState(GL10.GL_VERTEX_ARRAY);
//设置颜色数组
gl.glEnableClientState(GL10.GL_COLOR_ARRAY);
//颜色渲染 glColorPointer(int size, int type, int stride, Buffer pointer)
gl.glColorPointer(4, GL10.GL_FIXED, 0, colorBuffer);
//设置三角形顶点
gl.glVertexPointer(3, GL10.GL_FIXED, 0, triggerBuffer1);
//绘制三角形
gl.glDrawArrays(GL10.GL_TRIANGLES, 0, 3);
//关闭顶点设置
gl.glDisableClientState(GL10.GL_COLOR_ARRAY);
/* 渲染正方形 */
//设置当前色为蓝色 glColor4f(float red, float green, float blue,
```

```java
            float alpha)
            gl.glColor4f(0.0f, 1.0f, 1.0f, 0.5f);
            //重置当前的坐标原点到屏幕中心
            gl.glLoadIdentity();
            //左移1.5单位，并移入屏幕6.0
            gl.glTranslatef(1.5f, 0.0f, -6.0f);
            gl.glRotatef(rotateQuad, 1.0f, 0.0f, 0.0f);
            //设置和绘制有三维坐标矩形的4个顶点
            gl.glVertexPointer(3, GL10.GL_FIXED, 0, quaterBuffer1);
            gl.glDrawArrays(GL10.GL_TRIANGLE_STRIP, 0, 4);
            //关闭顶点数组设置功能
            gl.glDisableClientState(GL10.GL_VERTEX_ARRAY);
            rotateTri += 0.2f;
            rotateQuad -= 0.2f;
        }
        public void onSurfaceChanged(GL10 gl, int width, int height){
            // TODO Auto-generated method stub
            float ratio = (float) width / height;
            //设置OpenGL场景的大小
            gl.glViewport(0, 0, width, height);
            //设置投影矩阵
            gl.glMatrixMode(GL10.GL_PROJECTION);
            //重置投影矩阵
            gl.glLoadIdentity();
            //设置视窗的大小
            gl.glFrustumf(-ratio, ratio, -1, 1, 1, 10);
            //选择模型观察矩阵
            gl.glMatrixMode(GL10.GL_MODELVIEW);
            //重置当前的坐标原点到屏幕中心
            gl.glLoadIdentity();
        }
        public void onSurfaceCreated(GL10 gl, EGLConfig config){
            //启用阴影平滑
            gl.glShadeModel(GL10.GL_SMOOTH);
            //黑色背景
            gl.glClearColor(0, 0, 0, 0);
            //设置深度缓存
            gl.glClearDepthf(1.0f);
            //启用深度测试
            gl.glEnable(GL10.GL_DEPTH_TEST);
```

```
            //所作深度测试的类型
            gl.glDepthFunc(GL10.GL_LEQUAL);
            //通知系统对透视进行修正
            gl.glHint(GL10.GL_PERSPECTIVE_CORRECTION_HINT,
            GL10.GL_FASTEST);
        }
    }
```

运行结果如图 10-1 所示。

图 10-1 OpenGL 绘制多边形

10.4 图像旋转

图像旋转的思路很简单，即先画出图形，再用一个浮点数来表示旋转角度，浮点数是 OpenGL 里最为常用的类型之一，因为需要精确度。主要代码如下（完整源代码：Android Development\Chapter10\Section10_2\OpenGLRotate\OpenGL）：

```
        floatrotateTri;     //三角形旋转角度
```

然后启动旋转，通过 glRotatef 方法关联旋转图像角度：

```
    glRotatef(float angle, float x, float y, float z)
        gl.glRotatef(rotateTri, 0.0f, 1.0f, 0.0f);
```

最后不断改变旋转角度值，以实现不停旋转的效果：

```
            rotateTri -= 0.8f;
```

不用循环控制，因为已经重新定义了 onDrawFrame 函数。

运行结果如图 10-2 所示。

图 10-2 OpenGL 图形旋转

10.5 3D 三维实体空间

前面画的图形都是在一个面上画出来的，或者只有一个面，接下来看看真正的 3D 空间。还是以之前的三角形和四边形为基础，衍生为 3D 空间的四棱锥和立方体，主观上来讲，就是在空间上多出一个顶点来，把所有顶点连起来就构成了三维结构图像实体。

下面来看看具体的例子说明（完整源代码：Android Development\Chapter10\Section10_3\OpenGLRotate3D\Rotate3D_Activity）。

1. 空间实体的顶点设置

四棱锥的顶点是共享的，其他 4 个顶点分开在不同的位置，由 4 个三角形面组成，这样思考四棱锥就简单一些了。所有面的顶点都按逆时针顺序定义：

```
//定义四棱锥 4 个面的顶点
private IntBuffer triggerBuffer = BufferUtil.fBuffer(new int[]{
        0,one,0,
        -one,-one,0,
        one,-one,one,

        0,one,0,
        one,-one,one,
        one,-one,-one,

        0,one,0,
        one,-one,-one,
```

```
        -one,-one,-one,

        0,one,0,
        -one,-one,-one,
        -one,-one,one});
//定义立方体 6 个面的顶点
private IntBuffer quaterBuffer = BufferUtil.fBuffer(new int[]{
        one,one,-one,
        -one,one,-one,
        one,one,one,
        -one,one,one,

        one,-one,one,
        -one,-one,one,
        one,-one,-one,
        -one,-one,-one,

        one,one,one,
        -one,one,one,
        one,-one,one,
        -one,-one,one,

        one,-one,-one,
        -one,-one,-one,
        one,one,-one,
        -one,one,-one,

        -one,one,one,
        -one,one,-one,
        -one,-one,one,
        -one,-one,-one,

        one, one, -one,
        one, one, one,
        one, -one, -one,
        one, -one, one,
});
```

2．顶点颜色渲染

四棱锥由于顶点是同一个，所以颜色一样。代码如下：

```
//四棱锥顶点颜色渲染
private IntBuffer colorBuffer = BufferUtil.fBuffer(new int[]{
        one,0,0,one,
        0,one,0,one,
```

```
        0,0,one,one,

        one,0,0,one,
        0,one,0,one,
        0,0,one,one,

        one,0,0,one,
        0,one,0,one,
        0,0,one,one,

        one,0,0,one,
        0,one,0,one,
        0,0,one,one,
});
//立方体顶点颜色渲染
private IntBuffer colorBufferForQuad = BufferUtil.fBuffer(new int[]{
        0,one,0,one,
        0,one,0,one,
        0,one,0,one,
        0,one,0,one,

        one, one/2, 0, one,
        one, one/2, 0, one,
        one, one/2, 0, one,
        one, one/2, 0, one,

        one,0,0,one,
        one,0,0,one,
        one,0,0,one,
        one,0,0,one,

        one,one,0,one,
        one,one,0,one,
        one,one,0,one,
        one,one,0,one,

        0,0,one,one,
        0,0,one,one,
        0,0,one,one,
        0,0,one,one,

        one,0,one,one,
```

```
            one,0,one,one,
            one,0,one,one,
            one,0,one,one,
    });
```

3. 绘制三维实体图形

四棱锥有 4 个面，需要绘制 4 个平面图形来组合，用循环实现：

```
//绘制三角锥
for(int i=0; i<4; i++){
    gl.glDrawArrays(GL10.GL_TRIANGLE_STRIP, i*3, 3);
}
//绘制正方形
for(int i=0; i<6; i++){
    gl.glDrawArrays(GL10.GL_TRIANGLE_STRIP, i*4, 4);
}
```

运行结果如图 10-3 所示。

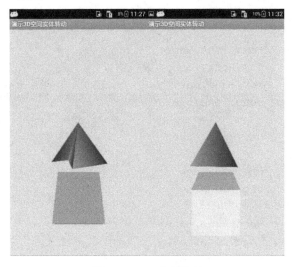

图 10-3　3D 立体旋转

10.6　映射纹理

我们在 3D 空间绘制了逼真的三维实体，但是在实际中如果只有单纯的颜色渲染就太单调了，不具有多样性与实用性。为了满足用户的视觉效果，可以在三维实体的表面显示图案或贴上用户自定义的各种图片，这种做法在游戏中是经常需要的。下面来看一个例子（完整源代码：AndroidDevelopment\Chapter10\ Section10_4\OpenGLTexture\TextureActivity）：

```
public class MyGLRender implements Renderer{
    float xrot, yrot, zrot;//三个坐标方向上的角度变化
```

```
int texture = -1;
int one = 0x10100;

IntBuffer vertices = BufferUtil.fBuffer(new int[]{
        -one,-one,one,
        one,-one,one,
        one,one,one,
        -one,one,one,

        -one,-one,-one,
        -one,one,-one,
        one,one,-one,
        one,-one,-one,

        -one,one,-one,
        -one,one,one,
        one,one,one,
        one,one,-one,

        -one,-one,-one,
        one,-one,-one,
        one,-one,one,
        -one,-one,one,

        one,-one,-one,
        one,one,-one,
        one,one,one,
        one,-one,one,

        -one,-one,-one,
        -one,-one,one,
        -one,one,one,
        -one,one,-one,
});
//纹理映射数据
IntBuffer texCoords = BufferUtil.fBuffer(new int[]{
    one,0,0,0,0,one,one,one,
    0,0,0,one,one,one,one,0,
    one,one,one,0,0,0,0,one,
    0,one,one,one,one,0,0,0,
    0,0,0,one,one,one,one,0,
    one,0,0,0,0,one,one,one,
});
//绘制纹理映射图形时使用的数组
FloatBuffer indices = BufferUtil.flBuffer(new float[]{
```

```
            0,1,3,2,
            4,5,7,6,
            8,9,11,10,
            12,13,15,14,
            16,17,19,18,
            20,21,23,22,
});
@Override
public void onDrawFrame(GL10 gl){
        //清除屏幕和深度缓存
    gl.glClear(GL10.GL_COLOR_BUFFER_BIT |
    GL10.GL_DEPTH_BUFFER_BIT);
        //重置当前坐标原点中心在屏幕中心
    gl.glLoadIdentity();
        //中心向里移动
    gl.glTranslatef(0.0f, 0.0f, -1.0f);
        //设置3个方向的旋转
    gl.glRotatef(xrot, 0.5f, 0.0f, 0.0f);
    gl.glRotatef(yrot, 0.0f, 0.5f, 0.0f);
    gl.glRotatef(zrot, 0.0f, 0.0f, 0.5f);
        //绑定纹理数组
    gl.glBindTexture(GL10.GL_TEXTURE_2D, texture);
        //开启纹理
    gl.glEnableClientState(GL10.GL_VERTEX_ARRAY);
    gl.glEnableClientState(GL10.GL_TEXTURE_COORD_ARRAY);
        //设置纹理和四边形对应的顶点
    gl.glVertexPointer(3, GL10.GL_FIXED, 0, vertices);
        //将纹理绑定到四边形上
    gl.glTexCoordPointer(2, GL10.GL_FIXED, 0, texCoords);
        //绘制纹理映射图形
    gl.glDrawElements(GL10.GL_TRIANGLE_STRIP, 50, GL10.GL_UNSIGNED_
    BYTE, indices);
        //设置完纹理后关闭
    gl.glDisableClientState(GL10.GL_TEXTURE_COORD_ARRAY);
    gl.glDisableClientState(GL10.GL_VERTEX_ARRAY);
        //三个坐标方向上的角度变化规律
    xrot+=0.3f;
    yrot+=0.3f;
    zrot+=0.3f;
}
@Override
public void onSurfaceChanged(GL10 gl, int width, int height){
    float ratio = (float) width / height;
        //设置OpenGL场景的大小
    gl.glViewport(0, 0, width, height);
        //设置投影矩阵
```

```
            gl.glMatrixMode(GL10.GL_PROJECTION);
            //重置投影矩阵
            gl.glLoadIdentity();
            //设置视窗的大小
            gl.glFrustumf(-ratio, ratio, -1, 1, 1, 1);
            //选择模型观察矩阵
            gl.glMatrixMode(GL10.GL_MODELVIEW);
            //重置当前坐标原点中心在屏幕中心
            gl.glLoadIdentity();
        }
        @Override
        public void onSurfaceCreated(GL10 gl, EGLConfig config){
            //白色背景
            gl.glClearColor(one, one, one, 0);
            gl.glEnable(GL10.GL_CULL_FACE);
            //启用阴影平滑
            gl.glShadeModel(GL10.GL_SMOOTH);
            //启用深度测试
            gl.glEnable(GL10.GL_DEPTH_TEST);
            //启用纹理映射
            gl.glClearDepthf(1.0f);
            //深度测试的类型
            gl.glDepthFunc(GL10.GL_LEQUAL);
            //精细的透视修正
            gl.glHint(GL10.GL_PERSPECTIVE_CORRECTION_HINT,
            GL10.GL_NICEST);
            //允许2D贴图纹理
            gl.glEnable(GL10.GL_TEXTURE_2D);
            //设置允许
            IntBuffer intBuffer = IntBuffer.allocate(1);
            //创建纹理
            gl.glGenTextures(1, intBuffer);
            texture = intBuffer.get();
            //设置要使用的纹理
            gl.glBindTexture(GL10.GL_TEXTURE_2D, texture);
            //使用GLUtil类中的静态方法texImage2D生成有图片的纹理
            GLUtils.texImage2D(GL10.GL_TEXTURE_2D, 0, GLImage.mBitmap, 0);
            //线形滤波
            gl.glTexParameterx(GL10.GL_TEXTURE_2D, GL10.GL_TEXTURE_MIN_
            FILTER, GL10.GL_LINEAR);
            gl.glTexParameterx(GL10.GL_TEXTURE_2D, GL10.GL_TEXTURE_MAG_
            FILTER, GL10.GL_LINEAR);
        }
    }
```

com/TextureActivity/TextureActivity.java 的内容如下所示：
```
    public class TextureActivity extends Activity{
```

```
Renderer render = new MyGLRender();
/** Called when the activity is first created. */
@TargetApi(Build.VERSION_CODES.CUPCAKE)
@Override
public void onCreate(Bundle savedInstanceState){
    super.onCreate(savedInstanceState);
    GLImage.load(this.getResources());//本地获取图片资源
    GLSurfaceView glView = new GLSurfaceView (this);

    glView.setRenderer(render);
    setContentView(glView);
    }
}
//定义一个类获取图片资源
class GLImage{
    public static Bitmap mBitmap;
    public static void load(Resources resources){
        mBitmap = BitmapFactory.decodeResource(resources, R.drawable.zw52);
    }
}
```

运行结果如图 10-4 所示。

图 10-4　映射纹理

10.7　光照与单击事件

Android OpenGL 的基本 3D 效果有时满足不了实际需要，如果添加一些人为的控制按键去操作图像，就会使 3D 效果更加逼真和美观了。下面就来学习一种人为的 3D 立方体操作（完整源代码：Android Development\Chapter10\Section10_5\OpenGL_Light_Event\

GLLightandeventActivity）：

com/GLLightandeventActivity/MyGLRender.java 的内容如下所示：

```java
public class MyGLRender implements Renderer{
    int one = 0x10000;
    float step = 0.4f;
    boolean key;
    boolean light = true;
    float xrot, yrot,zrot;  //X、Y、Z 轴旋转
    float xspeed, yspeed,zspeed;//旋转的速度
    //定义环境光(red,green,blue,alpha)
    FloatBuffer lightAmbient=FloatBuffer.wrap(new float[]{0.5f,0.5f,0.5f,
    1.0f});
    //定义漫射光,全部取 1.0f 表示最亮的漫射光
    FloatBuffer lightDiffuse=FloatBuffer.wrap(new float[]{1.0f,1.0f,1.0f,
    1.0f});
    //光源的坐标位置
    FloatBuffer lightPosition=FloatBuffer.wrap(new float[]{0.0f,0.0f,2.0f,
    1.0f});
    //纹理效果
    int [] texture;//[]里面的整数表示过滤的类型filter,可以为整数
    IntBuffer vertices = BufferUtil.fBuffer(new int[]{
        -one,-one,one,
        one,-one,one,
        one,one,one,
        -one,one,one,

        -one,-one,-one,
        -one,one,-one,
        one,one,-one,
        one,-one,-one,

        0,-one,0,
        0,-one,0,
        0,-one,0,
        0,-one,0,

        one,0,0,
        one,0,0,
        one,0,0,
        one,0,0,

        -one,0,0,
        -one,0,0,
        -one,0,0,
```

```java
        -one,0,0,
});
IntBuffer texCoords = BufferUtil.fBuffer(new int[]{
    one,0,0,0,0,one,one,one,
    0,0,0,one,one,one,one,0,
    one,one,one,0,0,0,0,one,
    0,one,one,one,0,0,0,0,
    0,0,0,one,one,one,one,0,
    one,0,0,0,0,one,one,one,
});
FloatBuffer indices = BufferUtil.fBuffer(new float[]{
    0,1,3,2,
    4,5,7,6,
    8,9,11,10,
    12,13,15,14,
    16,17,19,18,
    20,21,23,22,
});
@Override
public void onDrawFrame(GL10 gl){
    //清除屏幕和深度缓存
    gl.glClear(GL10.GL_COLOR_BUFFER_BIT |
    GL10.GL_DEPTH_BUFFER_BIT);
    //重置当前的原点为屏幕中心
    gl.glLoadIdentity();
    //启用GL光线设置，否则GL_LIGHTING光线就看不见
    gl.glEnable(GL10.GL_LIGHTING);
    //移动坐标中心
    gl.glTranslatef(0.0f, 2.0f, -5.0f);
    //设置X、Y、Z方向上的旋转
    gl.glRotatef(xrot, 1.0f, 0.0f, 0.0f);
    gl.glRotatef(yrot, 0.0f, 1.0f, 0.0f);
    gl.glRotatef(zrot, 0.0f, 0.0f, 1.0f);
    //选择使用的纹理
    gl.glBindTexture(GL10.GL_TEXTURE_2D, texture[1]);//texture[1]
    里面的1表示过滤类型设置为1
    gl.glNormalPointer(GL10.GL_FIXED, 0, normals);
    gl.glVertexPointer(3, GL10.GL_FIXED, 0, vertices);
    gl.glTexCoordPointer(2, GL10.GL_FIXED, 0, texCoords);
    gl.glEnableClientState(GL10.GL_NORMAL_ARRAY);
    gl.glEnableClientState(GL10.GL_VERTEX_ARRAY);
    gl.glEnableClientState(GL10.GL_TEXTURE_COORD_ARRAY);
    //绘制四边形
    gl.glDrawElements(GL10.GL_TRIANGLE_STRIP, 24, GL10.GL_UNSIGNED_
```

```
                BYTE, indices);
                gl.glDisableClientState(GL10.GL_TEXTURE_COORD_ARRAY);
                gl.glDisableClientState(GL10.GL_VERTEX_ARRAY);
                gl.glDisableClientState(GL10.GL_NORMAL_ARRAY);
                //旋转角度
                if ( key ){
                    xrot-=xspeed;
                    yrot-=yspeed;
                    zrot-=zspeed;
                }
                if (!light){         //判断是否开始光源（light==ture）
                    gl.glDisable(GL10.GL_LIGHT1);  // light!=ture 则禁用光源
                }else{           //light==ture
                    gl.glEnable(GL10.GL_LIGHT1);       //启用一号光源
                }
            }
            @Override
            public void onSurfaceChanged(GL10 gl, int width, int height){
                float ratio = (float) width / height;
                //设置OpenGL场景的大小
                gl.glViewport(0, 0, width, height);
                //设置投影矩阵
                gl.glMatrixMode(GL10.GL_PROJECTION);
                //重置投影矩阵
                gl.glLoadIdentity();
                //设置视窗的大小
                gl.glFrustumf(-ratio, ratio, -1, 1, 1, 10);
                //选择模型观察矩阵
                gl.glMatrixMode(GL10.GL_MODELVIEW);
                //重置模型观察矩阵
                gl.glLoadIdentity();
            }
            @Override
            public void onSurfaceCreated(GL10 gl, EGLConfig config){
                //告诉系统对透视进行修正
                gl.glHint(GL10.GL_PERSPECTIVE_CORRECTION_HINT, GL10.GL_FASTEST);
                //设置为黑色背景
                gl.glClearColor(0, 0, 0, 0);
                gl.glEnable(GL10.GL_CULL_FACE);
                //启用阴影平滑
                gl.glShadeModel(GL10.GL_SMOOTH);
                //启用深度测试
                gl.glEnable(GL10.GL_DEPTH_TEST);
                IntBuffer textureBuffer = IntBuffer.allocate(3);
```

```java
        //创建纹理
        gl.glGenTextures(3, textureBuffer);
        texture = textureBuffer.array();
        //创建 Nearest 滤波贴图
        gl.glBindTexture(GL10.GL_TEXTURE_2D, texture[0]);//过滤类型 0
        gl.glTexParameterx(GL10.GL_TEXTURE_2D,GL10.GL_TEXTURE_MAG_
        FILTER,GL10.GL_NEAREST); // ( NEW )
        gl.glTexParameterx(GL10.GL_TEXTURE_2D,GL10.GL_TEXTURE_MIN_
        FILTER,GL10.GL_NEAREST); // ( NEW )
        GLUtils.texImage2D(GL10.GL_TEXTURE_2D, 0, GLImage.mBitmap, 0);
        //创建线性滤波纹理
        gl.glBindTexture(GL10.GL_TEXTURE_2D, texture[1]);//过滤类型 1
        gl.glTexParameterx(GL10.GL_TEXTURE_2D,GL10.GL_TEXTURE_MAG_
        FILTER,GL10.GL_LINEAR); // ( NEW )
        gl.glTexParameterx(GL10.GL_TEXTURE_2D,GL10.GL_TEXTURE_MIN_
        FILTER,GL10.GL_LINEAR); // ( NEW )
        GLUtils.texImage2D(GL10.GL_TEXTURE_2D, 0, GLImage.mBitmap, 0);
        gl.glBindTexture(GL10.GL_TEXTURE_2D, texture[2]);//过滤类型 2
        gl.glTexParameterx(GL10.GL_TEXTURE_2D,GL10.GL_TEXTURE_MAG_
        FILTER,GL10.GL_NEAREST); // ( NEW )
        gl.glTexParameterx(GL10.GL_TEXTURE_2D,GL10.GL_TEXTURE_MIN_
        FILTER,GL10.GL_LINEAR); // ( NEW )
        GLUtils.texImage2D(GL10.GL_TEXTURE_2D, 0, GLImage.mBitmap, 0);
        gl.glClearDepthf(1.0f);
        gl.glDepthFunc(GL10.GL_LEQUAL);
        gl.glHint(GL10.GL_PERSPECTIVE_CORRECTION_HINT, GL10.GL_NICEST);
        gl.glEnable(GL10.GL_TEXTURE_2D);
        //设置环境光
        gl.glLightfv(GL10.GL_LIGHT1, GL10.GL_AMBIENT, lightAmbient);
        //设置漫射光
        gl.glLightfv(GL10.GL_LIGHT1, GL10.GL_DIFFUSE, lightDiffuse);
        //设置光源的位置
        gl.glLightfv(GL10.GL_LIGHT1, GL10.GL_POSITION, lightPosition);
        //启用一号光源
        gl.glEnable(GL10.GL_LIGHT1);
    }
    //单击按键事件
    public boolean onKeyDown(int keyCode, KeyEvent event){
        switch ( keyCode ){
            case KeyEvent.KEYCODE_DPAD_UP:
                key = true;
                xspeed=-step;
                zspeed=-step;
                break;
```

```java
                case KeyEvent.KEYCODE_DPAD_DOWN:
                    key = true;
                    xspeed=step;
                    zspeed=step;
                    break;
                case KeyEvent.KEYCODE_DPAD_LEFT:
                    key = true;
                    yspeed=-step;
                    zspeed=-step;
                    break;
                case KeyEvent.KEYCODE_DPAD_RIGHT:
                    key = true;
                    yspeed=step;
                    zspeed=step;
                    break;
                case KeyEvent.KEYCODE_DPAD_CENTER:
                    light = !light;
                    break;
            }
            return false;
        }
        public boolean onKeyUp(int keyCode, KeyEvent event){
            key = false;
            return false;
        }
    }
```

com/GLLightandeventActivity/GLLightandeventActivity.java 的内容如下所示：

```java
public class GLLightandeventActivity extends Activity{
    MyGLRender render = new MyGLRender();
    /** Called when the activity is first created. */
    @TargetApi(Build.VERSION_CODES.CUPCAKE)
    @Override
    public void onCreate(Bundle savedInstanceState){
        super.onCreate(savedInstanceState);
        GLImage.load(this.getResources());
        GLSurfaceView glView = new GLSurfaceView(this);
        glView.setRenderer(render);
        setContentView(glView);
    }
    //处理事件
    public boolean onKeyUp(int keyCode, KeyEvent event){
        render.onKeyUp(keyCode, event);
        return true;
```

```
        }
        public boolean onKeyDown(int keyCode, KeyEvent event)
        {
            render.onKeyDown(keyCode, event);
            return super.onKeyDown(keyCode, event);
        }
    }
    class GLImage{
        public static Bitmap mBitmap;
        public static void load(Resources resources){
            mBitmap = BitmapFactory.decodeResource(resources,
            R.drawable.image);
        }
    }
```

运行结果如图 10-5 所示。

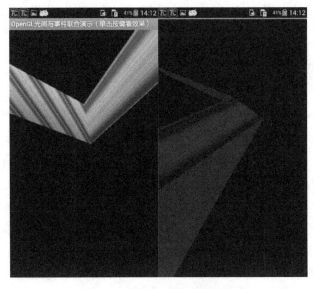

图 10-5　单击按键的光照效果

第 11 章
Android 传感器开发

随着手机的发展,现在各大手机支持的传感器类型越来越多,在开发中利用传感器进行某些操作令人们有一种耳目一新的感觉,如微信中的摇一摇,以及手机音乐播放器中的摇一摇切歌。本章简单介绍 Android 系统中传感器的使用及一些常用的传感器。

11.1 传感器种类

11.1.1 GPS

GPS 是英文 Global Positioning System(全球定位系统)的简称。GPS 起始于 1958 年美国军方的一个项目,1964 年投入使用。20 世纪 70 年代,美国陆海空三军联合研制了新一代卫星定位系统 GPS。GPS 由三部分组成:GPS 卫星组成的空间部分、若干地球站组成的控制部分和普通用户手中的接收机。对于手机用户来说,手机就是 GPS 定位系统的接收机。在 Android 应用开发中,程序员可以通过使用 Android 系统提供的 LocationManager 类及其他几个辅助类方便快捷地开发出 GPS 应用。Android 系统为 GPS 功能专门提供了一个 LocationManager 类,所有 GPS 定位相关的服务、对象都由该对象来生成。程序通过调用 Context 的 getSystemService()方法来获取该对象的实例。

11.1.2 动作传感器

Android 平台支持一些用于监视设备动作的传感器(这样的传感器共有 5 个)。其中两个[加速传感器(TYPE_ACCELEROMETER)和陀螺仪传感器(TYPE_GYROSCOPE)]是纯硬件传感器。另外三个(重力传感器、线性加速传感器和旋转向量传感器)既可能是硬件传感器,也可能是软件传感器。例如,在一些 Android 设备中,这些基于软件的传感器会从加速传感器和磁场传感器中获取数据,但在另一些 Android 设备中也可能从陀螺仪传感器中获取数据。也就是说,同一种基于软件的传感器在不同的 Android 设备中回传的数据可能来自不同的硬件传感器。因此，基于软件的同一种传感器在不同设备中的精确度、使用范围有所不同。大多数高端 Android 设备都有加速度传感器，还有一些拥有陀螺仪传感器。

11.1.3 位置传感器

Android 平台提供了两个传感器用于确定设备的位置，这两个传感器是磁场传感器和方向传感器。Android 平台还提供了测量设备正面到某一个邻近物体距离的传感器（邻近传感器）。磁场传感器和邻近传感器是基于硬件的传感器。大多数 Android 手机和平板电脑都有磁场传感器。而邻近传感器通常在手机中较为常见。因为可以用该传感器测试接听电话时手机屏幕离脸有多远。它可以在贴近耳朵接听电话时完成某些工作。方向传感器是基于软件的，该传感器的回传数据来自加速度传感器和磁场传感器。

11.1.4 环境传感器

Android 平台提供了 4 个传感器，用于检测不同的外部环境。我们可以使用这些传感器检测周围空气的湿度、光线、空气的压强和温度。这 4 个传感器都是基于硬件的传感器。除了光线传感器外，其他 3 个传感器在普通的 Android 设备中很少见。如果使用环境传感器，运行时最好检测当前 Android 设备是否支持该传感器。

11.2 GPS 应用

11.2.1 我的位置

Android 系统中的 LocationManager 为开发者提供了一系列方法来解决与 GPS 相关的问题，包括查询到上一个已知位置、注册或注销来自某个 LocationProvider 的周期性的位置更新，以及注册或注销在接近某个坐标时对一个已定义 Intent 的触发等。现在就以获取当前所在的位置为例介绍 Android 系统中 LocationManager 的简单使用。

首先需要获取 LocationManager 的一个实例，需要注意的是它的实例只能通过下面这种方式来获取，LocationManager 不允许直接实例化：

```
LocationManager LM = (LocationManager)getSystemService(Context.LOCATION_SERVICE)
```

得到了 LocationManager 的实例 LocationManager 以后，通过以下语句来注册一个周期性的位置更新：

```
locationManager.requestLocationUpdates(LocationManager.GPS_PROVIDER,
1000, 0, locationListener);
```

这句代码告诉系统，我们需要从 GPS 获取位置信息，每隔 1000ms 更新一次，并且不考虑位置的变化。最后一个参数是 LocationListener 的一个引用，必须要实现这个类。

位置服务有一个位置的数据来源称为 Provider，可以分为 NETWORK_PROVIDER 和 GPS_PROVIDER，其中 NETWORK_PROVIDER 使用基站和 WiFi 信号来决定用户的位置，更新速度更快、耗电量更少，但可能精确度较低，而 GPS_PROVIDER 使用 GPS 卫星进行定位，精度高但只能在室外使用，耗电量也更高。在使用过程中可以选择使用其中一个 Provider，也可以两个 Provider 同时使用。[47]

11.2.2 更新位置

位置检测更新与监听器实现：

```
locationManager.requestLocationUpdates(provider, 0, 0, locationListener);
private LocationListener locationListener = new LocationListener() {
    @Override
    public void onLocationChanged(Location location) {
        convertCoor(location);
        showLocation();
    }
    @Override
    public void onStatusChanged(String provider, int status, Bundle extras) {
    }
    @Override
    public void onProviderEnabled(String provider) {
        location = locationManager.getLastKnownLocation(provider);
    }

    @Override
    public void onProviderDisabled(String provider) {
    }
}
```

通过获取最佳 Provider 获取位置坐标（定位的关键）：

```
location = locationManager.getLastKnownLocation(provider);
private void getBestProvider() {
    Criteria criteria = new Criteria();
    criteria.setAccuracy(Criteria.ACCURACY_FINE);//高精度
    criteria.setAltitudeRequired(false);//无海拔要求
    criteria.setBearingRequired(false);//无方位要求
    criteria.setCostAllowed(true);//允许产生资费

    // 获取最佳服务对象
```

```
        provider = locationManager.getBestProvider(criteria,true);
    }
```
由于地图所用的位置坐标标准不同于直接获取的位置坐标,所以需要进行坐标的转换:
```
    private void convertCoor(Location location) {
        //将GPS设备采集的原始GPS坐标转换成百度坐标
        CoordinateConverter converter = new CoordinateConverter();
        converter.from(CoordinateConverter.CoordType.GPS);
        converter.coord(new LatLng(location.getLatitude(),
        location.getLongitude()));
        desLatLng = converter.convert();
    }
```
最后是添加定位图标及显示位置:
```
    private void init() {
        Bitmap bitmap = Bitmap.createScaledBitmap(BitmapFactory.decode
        Resource(getResources(), R.mipmap.pointer), 100, 100, true);
        BitmapDescriptor bitmapDescriptor = BitmapDescriptorFactory.
        fromBitmap(bitmap);

        baiduMap.setMyLocationEnabled(true);
        MyLocationConfiguration configuration = new MyLocationConfiguration
        (MyLocationConfiguration.LocationMode.NORMAL, true, bitmapDescriptor);

        baiduMap.setMyLocationConfigeration(configuration);
        if (isFirst) {
            MapStatusUpdate update = MapStatusUpdateFactory.newLatLng
                (desLatLng);
            baiduMap.animateMapStatus(update);
            update = MapStatusUpdateFactory.zoomBy(5f);
            baiduMap.animateMapStatus(update);
            isFirst = false;
            createDialog();
        }
    }
    private void showLocation() {
        if (desLatLng != null) {
            MyLocationData.Builder data = new MyLocationData.Builder();
            data.latitude(desLatLng.latitude);
            data.longitude(desLatLng.longitude);
            data.direction(currentRotation);
            baiduMap.setMyLocationData(data.build());
        }
    }
```

11.2.3 地图功能

1．MapController

MapController 可以控制地图的移动和伸缩。基本上是以鼠标所在的 GPS 坐标为中心来控制 MapView 中的 View 组件，管理 Overlay，同时提供视图的基本功能。利用多种视图模式［地图模式（某些城市可实时对交通状况进行更新）、卫星模式、街景模式］来查看 Google Map。

常用方法：animateTo（GeoPoint point）、setCenter（GeoPoint point）、setZoom（int zoomLevel）。

2．MapView

MapView 是用来显示地图的 View，它派生自 android.view.ViewGroup。当 MapView 获得焦点后，可以控制地图的移动和缩放。

地图可以通过不同的形式显示出来，如街景模式、卫星模式等，通过 setSatellite（boolean）、setTraffic（boolean）、setStreetView（boolean）方法来实现。

MapView 只能被 MapActivity 创建，这是因为 MapView 需要通过后台的线程来连接网络或文件系统，而这些线程要由 MapActivity 来管理。[48]

常用方法：getController()、getOverlays()、setSatellite（boolean）、setTraffic（boolean）、setStreetView（boolean）、setBuiltInZoomControls（boolean）等。

3．MapActivity

管理 Activity 的生命周期，为 MapView 建立及取消对 Map Service 的连接。

MapActivity 是一个抽象类，任何想要显示 MapView 的 Activity 都需要派生自 MapActivity。并且在其派生类的 onCreate()中，都要创建一个 MapView 实例。可以通过 MapViewconstructor［然后添加到 View 中的 ViewGroup.addView（View）］或 layout XML 来创建。[49]

4．Overlay

Overlay 覆盖到 MapView 的最上层，可以扩展其 ondraw 接口，自定义在 MapView 中显示一些内容。MapView 通过 MapView.getOverlays()对 Overlay 进行管理。

除了 Overlay 这个基类外，Google 还扩展了两个比较有用的 Overlay。

（1）MylocationOverlay：集成了 Android.location 中接收当前坐标的接口，集成了 SersorManager 中 CompassSensor 的接口。

只需要使用 enableMyLocation()，enableCompass 就可以让程序拥有实时的 MyLocation 及 Compass 功能（Activity.onResume()中）。

（2）ItemlizedOverlay：管理一个 OverlayItem 链表，用图片等资源在地图上进行风格相同的标记。[49]

5．Projection

Projection 的类如表 11-1 所示。

表 11-1　Projection 的类

GeoPoint	fromPixels(int x, int y) 从像素坐标来创建一个新的 GeoPoint，坐标原点是 MapView 的左上角
float	metersToEquatorPixels(float meters) 在当前所用缩放级别下，将以米为单位的距离（沿赤道）转换为以像素为单位的距离（水平）
android.graphics.Point	toPixels(GeoPoint in, android.graphics.Point out) 将给定的 GeoPoint 转换为屏幕上的像素坐标，坐标原点是 MapView 的左上角

11.3　Acceleration 传感器

传感器的存在和发展，让物体有了触觉、味觉和嗅觉等感官，让物体慢慢变得"活"了起来。当前的 Android 设备中已经集成了数十个传感器，比较常见的有加速度传感器、陀螺仪传感器、邻近传感器等。虽然种类繁多，但在 Framework 中仅仅提供了几个类和接口就实现了传感器的相关功能。

从 Acceleration 传感器获得数据示例的代码如下：

```
public void onSensorChanged(SensorEvent sensorEvent) {
    if (sensorEvent.sensor.getType() == Sensor.TYPE_ACCELEROMETER){
        accelerometerValues = sensorEvent.values.clone();
    }else if (sensorEvent.sensor.getType() == Sensor.TYPE_MAGNETIC_FIELD){
        maneticValues = sensorEvent.values.clone();
    }
    float[] R = new float[9];
    float[] values = new float[3];
    SensorManager.getRotationMatrix(R,null,accelerometerValues,maneticValues);
    SensorManager.getOrientation(R,values);
    float rotateDegree = -(float) Math.toDegrees(values[0]);
    if (Math.abs(rotateDegree - lastDegree) > 1){
        RotateAnimation animation = new RotateAnimation(lastDegree,
        rotateDegree, Animation.RELATIVE_TO_SELF,0.5f,Animation.
        RELATIVE_TO_SELF,0.5f);
        animation.setFillAfter(true);
        compass.startAnimation(animation);
        lastDegree = rotateDegree;
    }
}
```

11.4 Gyroscope 传感器

Gyroscope 传感器就是手机内部的一个陀螺仪，它的轴由于陀螺效应始终与初始方向平行，这样就可以通过与初始方向的偏差计算出实际方向。手机里的陀螺仪实际上是一个结构非常精密的芯片，其内部包含超微小的陀螺。

陀螺仪测量时的参考标准是内部中间在与地面垂直的方向上进行转动的陀螺。通过设备与陀螺的夹角可得到结果。陀螺仪的强项在于测量设备自身的旋转运动。它对于设备的自身运动更擅长，但不能确定设备的方位。

陀螺仪对设备旋转角度的检测是瞬时的而且是非常精确的，能满足一些需要高分辨率和快速反应的应用的需求，如 FPS 游戏的瞄准。而且陀螺仪配合加速计可以在没有卫星和网络的情况下进行导航，这是陀螺仪的经典应用。同时处理直线运动和旋转运动时，就需要把加速计和陀螺仪结合起来使用。如果还想设备在运动时不至于迷失方向，需再加上磁力计。陀螺仪的 X、Y、Z 分别代表设备围绕 X、Y、Z 三个轴旋转的角速度：radians/s。因此，需要利用角速度与时间积分计算角度，得到的角度变化量与初始角度相加，就得到目标角度，其中积分时间 Δt 越小，输出角度越准，但陀螺仪的原理决定了它的测量基准是自身，并没有系统外的绝对参照物，加上 Δt 不可能无限小，则积分的累积误差会随着时间流逝迅速增加，最终导致输出角度与实际不符。因此，陀螺仪只能工作在相对较短的时间尺度内。[50] 代码如下所示：

```
private static final float NS2S = 1.0f / 1000000000.0f;
private float timestamp;
public void onSensorChanged(SensorEvent event){
    if (timestamp != 0) {
        final float dT = (event.timestamp - timestamp) * NS2S;
        angle[0] += event.data[0] * dT;
        angle[1] += event.data[1] * dT;
        angle[2] += event.data[2] * dT;
    }
    timestamp = event.timestamp;
}
```

11.5 Proximity 传感器

Proximity 传感器就是距离传感器，又称为位移传感器，是传感器的一种，用于感应其与某物体间的距离以完成预设的某种功能，目前已得到相当广泛的应用。

距离传感器的原理：利用各种元件检测对象物的物理变化量，通过将该变化量换算为距离，来测量从传感器到对象物的距离位移。根据使用元件不同，距离传感器分为光学式位移传感器、线性接近传感器、超声波位移传感器等。手机中使用的距离传感器是利用测时间来实现距离测量的一种传感器。红外脉冲传感器发射特别短的光脉冲，通过测量此光脉冲从发射到被物体反射回来的时间，可以计算出手机与物体之间的距离。[51]

从 Proximity 传感器获得数据示例的代码如下：

```
public void onSensorChanged(SensorEvent event) {
    if (event.sensor.getType() != Sensor.TYPE_PROXIMITY)  return;
    ong curTime = System.currentTimeMillis();
    if (lastUpdate == -1 || (curTime - lastUpdate) > mCycle) {
        lastUpdate = curTime;
        float lastValue = value;
        value = event.values[SensorManager.DATA_X];
        if (lastEvent == -1 || (curTime - lastEvent) > mEventCycle) {
            if (Math.abs(value - lastValue) > mAccuracy) {
                lastEvent = curTime;
                mBrowser.eventList.run(EVENT_ONPROXIMITYCHANGED);
            }
        }
    }
}
```

第 12 章 Android NDK 开发技术

Native Development Kit，简称 NDK，是一种基于原生程序接口的软件开发工具。通过此工具开发的程序直接以本地语言运行，而非虚拟机。因此，只有 Java 等基于虚拟机运行的语言的程序才会有原生开发工具包。NDK 是一系列工具的集合，可帮助开发者快速开发 C（或 C++）的动态库，并能自动将 So 和 Java 应用一起打包成 APK。这些工具对开发者的帮助是巨大的。NDK 集成了交叉编译器，并提供了相应的.mk 文件来屏蔽 CPU、平台、ABI 等差异，开发人员只需要简单修改.mk 文件（指出"哪些文件需要编译"、"编译特性要求"等），就可以创建出 So。NDK 可以自动地将 So 和 Java 应用一起打包，极大地减轻了开发人员的打包工作。本章讲述 NDK 的基本使用过程。

12.1 NDK 环境的搭建

开发工具：Android Studio 2.1.2。
NDK 版本：android-ndk-r10e，支持 64 位 So 库的编译。
JDK 版本：1.8，64 位。
下载 NDK：可以通过 Android Studio 的 SDK Manager 下载，并且可以根据自己的计算机系统自行选择最新版本。

12.2 新建 NDK 工程

（1）新建一个 Project，开始 NDK 的配置。
（2）在工程的 MainActivity 里添加如下代码：
```
static {
    System.loadLibrary("MyJni");//导入生成的链接库文件
```

```
    }
    public native String getStringFromNative();//本地方法
    public native String getString_From_c();
        abiFilters "armeabi","armeabi-v7a","x86"
    }
```

（3）在 build.gradle 中的 defaultConfig{}标记里添加：

```
ndk{
    moduleName "MyJni"
}
```

（4）在 Project 工程下单击"Build→MakeProject"，生成.class 文件。

（5）通过 Terminal 中的命令来生成.h 文件：

```
ndk{
    moduleName "MyJni"
    ldLibs "log"
```

输入：cdapp\build\intermediates\classes\debug+Enter。

再输入：javah -classpath . -jnicom.Mainactivity。

此时，在 debug 目录下就会生成 com_ndkapp_Mainactuvity.h 文件。

（6）在 main 文件夹下建 jni 文件夹，将.h 头文件剪切到此文件夹中，并建立一个.c 文件，内容如下：

```
#include "com_ndkapp10_BuildConfig.h"
JNIEXPORT jstring JNICALL Java_com_ndkapp_MainActivity_getStringFromNative
(JNIEnv * env, jobject obj){
    return (*env)->NewStringUTF(env,"My name is Ouyangshengduo,Hi!");
    //return env->NewStringUTF("My name is Ouyangshengduo,Hi!");
    //c++调用
}
    JNIEXPORT jstring JNICALL Java_com_ndkapp_MainActivity_getString_1From_1c
    (JNIEnv * env, jobject jobject){
        return (*(*env)).NewStringUTF(env,"My name is");
        //return env->NewStringUTF("My name is Ouyangshengduo,Hi!");
        //c++调用
    }
```

（7）在 gradle.properties 文件中加入"android.useDeprecatedNdk=true"。

（8）在 Project 工程下单击"Build→Make Project"。

（9）在 build.gradle 中的 android{}标记里添加：

```
sourceSets.main{
```

```
            jni.srcDirs = []
            jniLibs.srcDir "src/main/libs"
    }
```
（10）在"app→build→intermediates→ndk→debug"下会生成.mk文件，将此文件剪切到jni文件夹中。

（11）用鼠标右键单击"mk→External Tools→ndk-build"生成.so文件。

（12）配置完成，运行MainActivity。

参考文献

[1] 玩转 Android 手机-百度文库互联网文档资源[EB].https://wenku.baidu.com/view /1ea7552db4daa58da0-114a83.html.2012.

[2] Android 的历史版本与开发博客频道-CSDN.NET[EB].http://blog.csdn.net/mozart_cai/article /details/18794817.2017.

[3] Android 历史版本[EB]．http://blog.csdn.net/haima1998/article/details/25886787.2014．

[4] Android 初学者入门[EB]．https://wenku.baidu.com/view/ecb367b8960590c69ec376 a1.html.2012．

[5] 3g 智能手机操作系统的研究和分析[EB]．http://www.doc88.com/p-093936800 4519 .html．2007.

[6] 三种嵌入式操作系统的分析与比较（2）[EB]．http://www.eepw.com.cn/article/3929 1.htm．2007．

[7] Object-Oriented Programming-百度百科[EB]．https://baike.baidu.com/item/Object-Oriented%20Programming/6979668．2012．

[8] 面向对象设计的三个基本要素与五个基本设计原则[EB]．https://www.cn blogs.com/zharma/p/4554188.html．2012．

[9] 面向对象系统分析与开发专题<1>[EB]．https://www.cnblogs.com/kingmoon /archive/2011/04/19/2020756.html．2011．

[10] 常见的程序设计方法及适用情况[EB]．https://wenku.baidu.com/view/109d11810c2259 0102029de5. html．2014．

[11] Java 语言特点[EB]．http://blog.csdn.net/lsj960922/article/details/79150534．2017．

[12] 高立军．有关 Java 语言的 Android 手机软件开发的分析[J].学园，2015（3）．

[13] XML 基础概念[EB]．https://www.cnblogs.com/zhengcheng/p/4278764.html．2015-02-07．

[14] 编辑器编译器与集成开发环境(IDE)[EB].http://blog.csdn.net/hyl52101314/article/details/53018212．2015-02-07．

[15] Eclipse 基础应用实例[EB]．https://wenku.baidu.com/view/53840830a32d7375a41780a9.html.2012．

[16] Android Studio IDEA：基于 IDEA 的 Android 开发环境[EB]．http://blog.csdn.net/phodal/article/details/8937229．2013-05-16．

[17] Android 模拟器使用模拟 SD 卡[EB]．http://blog.csdn.net/terryzero/article/details 5714917．2010-07-06．

[18] 李明浩．Traceview 的使用[EB]．http://blog.csdn.net/hudashi/article/details/70316702013．

[19] 安装 JDK 并配置环境变量[EB]．http://product.pconline.com.cn/itbk/software/dnwt/1408/5260596. html2017-07-15．

[20] Android 开发工具之 DDMS[EB]．http://www.jizhuomi.com/android/environment/82.html.2013．

[21] 锵鹏鹏. Android：DDMS 工具的使用[EB]．http://blog.sina.com.cn/s/blog_90551e38010118v3.html.2013．

[22] android-DDMS 工具[EB]．http://www.jb51.net/article/36668.html.2013．

[23] 什么是 NDK[EB].http://blog.csdn.net/mnorst/article/details/7026130．2011．

[24] Android 应用程序的五大基本组件 Activity[EB]．https://www.aliyun.com/jiaocheng/54789.html.2018．

[25] Activity 四种加载模式[EB]．https://jingyan.baidu.com/article/a378c960b33431b32928307f.html.2015．

[26] Android——Intent 简介[EB]．http://blog.csdn.net/qq_31370269/article/details/50725701.2016-02-23．

[27] Android 中的广播 Broadcast 详解[EB]．https://www.toutiao.com/i6437725990243271170/2017．

[28] Binder 机制和 AIDL 使用介绍[EB]．http://blog.csdn.net/gjr9596/article/details/52201400.2016．

[29] AndroidManifest.xml 中一些常用的属性[EB]．http://blog.chinaunix.net/uid-9185047-id-3460344.html. 2013．

[30] Android 中的 Surface 和 SurfaceView[EB]．https://www.cnblogs.com/Sharley/p/5600314.html.2016-06-20．

[31] Android Paint 和 Color 类[EB]．http://www.cnblogs.com/-OYK/archive/2011/10/25/2223624.html. 2016-03-02．

[32] Android UI 开发[EB]．http://blog.csdn.net/droidpioneer/article/details/24845043.2016．

[33] Code Library Free Source Code, Program Tips and Technology Documents[EB]．https://www.ucosoft.com/. 2013-03-06．

[34] Hello Android 学习之 SQLite[EB]．http://blog.csdn.net/zyhuangan/article/details/6985710.2013-03-06．

[35] Android 的多媒体框架 OpenCore 介绍|IT168 技术开发[EB]．http://www.360doc.com/content/10/0207/22/155970_15398760.shtml.2010．

[36] Android MediaPlayer 的生命周期[EB]．http://blog.csdn.net/haelang/article/details/43489833.2015．

[37] 邵艳洁．Android 操作系统移植及应用研究[D]．湖南：湖南大学，2011．

[38] android 开发入门音频视频开发[EB]．http://blog.51cto.com/ticktick/1956269.2017．

[39] 杨鹏飞，苗忠良．智能手机 camera 应用的设计与实现[J]．电脑知识与技术，2008（13）．

[40] 余朋．网络通信协议的分析与实现[J].电脑编程技巧与维护，2014(14)．

[41] Android 网络编程之 Http 通信[EB]．http://lib.csdn.net/article/android/7229.2017．

[42] 卡尔弗特，多纳霍. Java TCP/IP Socket 编程[M]．周恒民译. 北京：机械工业出版社，2017．

[43] Http 通信与 Socket 通信[EB]．https://www.cnblogs.com/jmqm/p/6701804.html.2016．

[44] Socket 与 ServerSocket 类介绍[EB]．https://www.cnblogs.com/cst11021/articles/4679737.html.2015．

[45] 吴海青．基于 Webkit 内核的手机浏览器的设计与实现[D]．北京：北京邮电大学，2011．

[46] 冯贤全．基于 Android 和 OpenGL 的多媒体播放器研究[D]．四川：电子科技大学，2012-03-01．

[47] Android 开发—百度地图开发[EB]．https://cnblogs.com/zhangmiao14/p/7274977.html.2017．

[48] 高柏俊．基于 LBS 的城市智能泊车系统研究与设计[D]．西安：电子科技大学，2017．

[49] 谷歌地图开发[EB]．https://wenku.baidu.com/view/150905bac77da26925c5b0d1.html2012．

[50] 陀螺仪、加速计、磁力计等传感器汇总[EB]．http://blog.csdn.net/a345017062/article/details/6459643. 2013．

[51] 穆冬梅．基于情境感知的移动终端推荐系统研究[D]．北京：北京邮电大学，2013．

[52] Android Studio 之 NDK 篇[EB]．https://www.cnblogs.com/fnlingnzb-learner/p/7207468.html. 2017．

[53] 使用 Android Studio 创建第一个 Hello World 应用程序[EB].http://blog.csdn.net/zdw_wym/article/details/49864673.2015．

[54] Android Studio 之 NDK 开发基础篇[EB].http://www.360doc.com/content/18/0203/02/52553602_727334006.shtml.2016．

[55] 尚学堂 Android 核心基础汇总[EB].http://www.9299.net/read/jj098jg625i52ii58gj6hk46.html.2017．

[56] Android 网络编程（一）HTTP 协议原理[EB].http://liuwangshu.cn/application/network/1-http.html.2016.
[57] Android 传感器的使用总结[EB].http://blog.csdn.net/wenzhi20102321/article/details/53282313.2016.
[58] Android GPS 应用开发[EB].http://blog.csdn.net/smilehanbright/article/details/72792733.2017.
[59] Android file 类使用详解-SDcard[EB].http://blog.csdn.net/codefarmercxy/article/details/54982162.2017.
[60] Android 数据存储——文件读写操作（File）[EB].https://www.cnblogs.com/LiHuiGe8/p/5604725.html. 2016.
[61] Android 总结：ContentProvider 的使用[EB].http://blog.csdn.net/u014136472/article/details/49907713. 2015.